THE · COMPLETE · BOOK · OF
HOME
PRESERVING

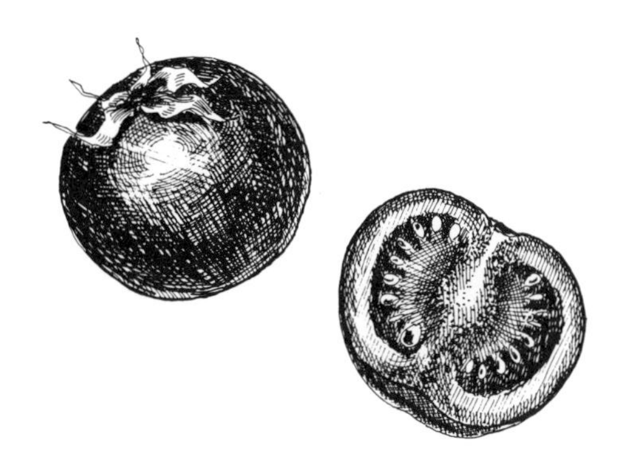

THE · COMPLETE · BOOK · OF HOME PRESERVING

Mary Norwak

WARD LOCK LIMITED
LONDON

First edition published by Ward Lock Limited 1978

This edition first published in Great Britain in 1988
by Ward Lock Limited, 8 Clifford Street
London W1X 1RB, an Egmont Company

House editor Barbara Fuller
Designed by Anita Ruddell
Cover photography by Jhon Kevern
Line illustrations by Antonia Enthoven
Photographs; Anthony Blake Photo Library pp 17, 32 below, 33, 65; Michael Boys Syndication pp 64, 96, 113; British Sugar Bureau p 16; Creda Microwaves Ltd p 129; Julia Hedgecoe pp 32 above, 144; Jhon Kevern p 145 and jacket photograph; Peter Myers pp 48, 49, 97, 112, 128; Silver Spoon Sugars p 48.

Text set in Goudy Old Style
by MS Filmsetting Limited, Frome, Somerset

Printed and bound in Great Britain by
Hazell, Watson & Viney Ltd,
Member of the BPCC Group,
Aylesbury, Bucks

British Library Cataloguing in Publication Data

Norwak, Mary, 1929–
The complete book of home preserving.
rev and updated ed.
1. Preserved foods. Recipes
I. Title II. Series
641.6′1

ISBN 0 7063 6718 9

WI MARKETS
The recipes and method of presentation in this book do not necessarily conform to the statutory requirements for selling in WI Markets, or for WI exhibiting and judging.

WI Life & Leisure Series
Home-made Wines, Syrups & Cordials 0 7063 67170

LIST OF CONTENTS

INTRODUCTION

Once upon a time, most houses had their stillroom and larder. The mistress of the house was in charge of the preserving of food and, with the help of her servants, would be responsible for drying herbs, fruit and vegetables; candying and crystallizing fruit and flowers; potting jams and preparing pickles, chutneys and sauces and salting or potting meat. A country household was totally self-sufficient, but preserving was necessary to ensure food supplies right through a year, when animals might have to be killed because of lack of fodder and every scrap of food was precious.

As commercial food processing became more sophisticated, the need for preserving food at home became less essential, although it enjoyed an enthusiastic revival during two World Wars. The development of the home freezer stimulated families to start preserving again by a simple method, which has led to enthusiasm for vegetable and fruit growing and for keeping poultry and in some cases, pigs. This enthusiasm has led in turn to an interest in other methods of preserving to give variety to meals, so that once again, we are experiencing a revival of interest in making jam and pickles, herb-growing, meat-salting and smoking.

This book is written with the intention of helping those who have not been used to home food preserving to tackle the various jobs simply and with pleasure, and I have carried them all out at home with the most basic equipment.

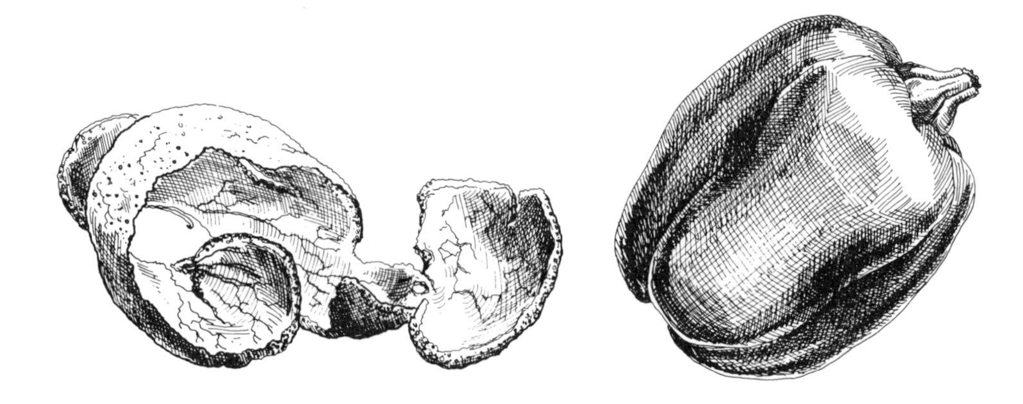

FRUIT BOTTLING

Success in bottling depends on three things: heating the fruit sufficiently; closing the bottles while they are hot; and making sure that airtight jars are used.

Some fruits need more heating than others, and if the fruit is tightly packed in the jar, so that there is less than 130ml/¼ pint syrup or water in a 500g/1lb jar, it needs longer heating than a looser pack. Large jars also need more heating than small ones. If the jars of fruit are heated in an oven, the time of heating will depend both on the kind of fruit used and on the number of jars being bottled.

• EQUIPMENT •

The preserving jars may have glass or metal lids which rest on a rubber band. Either clips or screw-bands are used to hold the lids tightly in place while the jars are cooling. When the jar is cold, the vacuum formed inside holds the lid in place.

Always examine the jars carefully before use, as any ridges or chips on the mouth of the jar or lid will prevent it from sealing. Wash all bottles thoroughly, rinse in clean hot water and invert them to drain. The rubber bands must fit properly and be soft and flexible. They should be soaked in boiling water just before they are put on the jar.

• PREPARATION OF THE FRUIT •

Soft fruit. Most berry fruits are best when they are just ripe, but not over-ripe, and as fresh as possible. Any unsound fruit, stalks or leaves should be removed and the fruit rinsed in clean, cold water.

Gooseberries bottle best when under-ripe. If preserved in syrup, cut a small slice off both ends of the berries when they are topped and tailed, or else prick the fruit, to prevent shrivelling.

Raspberries should be examined to see that they are free from maggots, but are better not rinsed.

Rhubarb is treated like soft fruit for bottling. Young spring rhubarb is the best, cut into short lengths (2.5–5cm/1–2in).

Strawberries lose colour and shrink considerably during bottling. This can be prevented to some extent if they are put in a basin, just covered with boiling sugar syrup (coloured with strawberry colour) and left overnight before they are packed into jars. This pack needs longer cooking than the unsoaked fruit.

Stone fruit. Stone fruit should be just ripe and rinsed in clean, cold water after removing stems or blemished fruit. Large fruits may be halved and the stones removed before bottling. Peaches should be dipped for 10–30 seconds in boiling water, so that the skins can be removed before bottling. Peaches and other fruits with light-coloured flesh should be covered with syrup or water as soon as possible after peeling to prevent discoloration.

Hard fruit. Apples and pears are generally peeled, cored and cut into slices or halves. If the sliced fruit is kept in water with 15g/½oz salt per 1 litre/2 pints during preparation, discoloration is retarded.

More apples can be packed into a bottle if the slices are put, a few at a time, into boiling water for 2–3 minutes until they are pliable, but not too soft. This is termed a 'solid pack' and needs longer cooking.

Tomatoes. Small whole tomatoes may be bottled and covered with brine made from 15g/½oz salt to 1 litre/2 pints water.

'Solid-pack tomatoes' are prepared by dipping the tomatoes in boiling water for 10–20 seconds, removing the outer skin, cutting large fruits into halves or quarters and packing them tightly in a jar with about 2 tsp salt and 1 tsp sugar to each 1kg/2lb fruit. Add ¼ tsp citric acid or 2 tsp lemon juice to each 500g/1lb tomatoes as the extra acid ensures safe bottling.

• Syrup

Fruit may be bottled in water or syrup. Fruit bottled in syrup keeps a better flavour and is ready for use when the bottles are opened, but it may rise a little in the bottles.

The quantity of sugar can be varied. A suitable strength for most fruits is made from 500g/1lb sugar to 1 litre/2 pints water, sufficient for about eight 500g/1lb jars. This should be boiled for 1 minute; it may be needed hot or cold according to the method used.

• CHOICE OF BOTTLING METHOD •

The water-bath Methods I and II are strongly recommended for reliable results, but necessitate using a deep pan. Method III is a quick one, but care in timing is needed otherwise the fruit may be over-cooked.

The oven methods are suitable only if the temperatures can be regulated at 130°C/250°F/gas mark ½ or 150°C/300°F/gas mark 2. They are more extravagant on fuel than the other methods. Method IV should not be used for fruits liable to discolour in the air, or for solid-pack tomatoes, apples or strawberries.

The pulping method is suitable for apples, tomatoes or other fruits that do not need to be kept whole.

Slow water bath: Method I. Any deep vessel, such as a fish-kettle, fitted with a false bottom, can be used, but the bottles must not be in direct contact with the base of the pan. Fill the bottles with fruit and cold syrup or water, put on the rings, lids and clips. If screw-bands are used, these are fitted on but unscrewed by a quarter turn during processing. Have sufficient cold water to cover the jars in the pan, and raise the temperature gradually to reach that given in the chart (see pages 12–13) for 1½ hours. Keep it at that temperature for the required time. The bottles should then be removed and placed on a wooden surface, the screw-bands tightened and left to cool.

Quick water bath: Method II. The bottles of fruit and hot syrup or water should be put in the pan as in Method I, but the water in the pan should be warm and raised to simmering point in 25–30 minutes. The jars are kept in the simmering water for the time given in the chart, then removed for cooling as in Method I.

Pressure cooker: Method III. A pressure pan fitted with a gauge or weight for 5lb/2.5kg (low) pressure is needed for fruits. The amount of water needed will vary with the size of pan to give 2.5cm/1in depth. The fruit is packed into warm jars, and boiling liquid added to within 2.5cm/1in of the top. The rings, lids and clips are fitted (screw-bands loosened by a quarter turn), and the jars put on a rack in the hot pan, while the lid is placed on with the vent open and heated until the steam appears; the vent is then closed and pressure brought up to 5lb/2.5kg (low). The time taken from placing on the lid to reaching pressure should be between 5 and 10 minutes. Maintain pressure for the time stated in the chart, then remove the pan from the heat and leave it to cool for 10 minutes in the air before opening. Then remove the bottles, tighten any screw-bands, and leave the jars on a wooden surface to cool.

Slow oven: Method IV (liquid added after heating fruit). This method can be used for green or dark-coloured fruits, but light ones discolour. The temperature reached is not high enough for solid-pack tomatoes, apples or soaked strawberries.

Pre-heat the oven for 15 minutes at a setting to give 130°C/250°F/gas mark ½ at the end of that time. As settings for the oven temperatures vary, an oven thermometer is the surest guide. Keep the oven at this setting during processing.

Pack the jars tightly with the prepared fruit, put on the lids, but do not add the syrup, rubber rings, clips or screw-bands. Put the jars 5cm/2in apart on cardboard or on a baking sheet lined with newspaper in the central part of the warm oven. Vary the cooking time with the kind of fruit and the total capacity of the jars in the oven, e.g. two 1kg/2lb jars require the same time as four 500g/1lb jars. After heating for the required time, remove the jars one at a time from the oven and fill each one quickly to the brim with boiling syrup or water, or for whole tomatoes, with boiling brine. Have ready the rubber rings in boiling water, quickly place a ring on the jar, replace the hot lid and fit on the clips or tightly screw on the screw-bands. Seal the jars with minimum delay after taking from the oven.

Moderate oven: Method V (liquid added before heating fruit). The oven should be set to give a temperature of 150°C/300°F/gas mark 2 after heating for 15 minutes (an oven thermometer is recommended) and this setting is used during processing.

Pack the fruit into warmed jars, fill to within 2.5cm/1in of the top with boiling syrup or water and adjust the rubber rings and lids. Screw-bands should not be put on until after processing. Position the jars 5cm/2in apart on a baking sheet lined with newspaper, and place it in the central part of the warm oven. Process for the time given in the chart, calculating the capacity of the jars as in Method IV. As soon as the jars are taken from the oven, put on any screw-bands and tighten at once.

◆ Pulping

This method is useful if whole fruit is not needed or if sieved fruit is required.

Stew the prepared fruit in a saucepan with enough water to prevent it from burning, and when cooked and boiling vigorously, pour quickly into hot, clean preserving jars and seal immediately with hot lids and rings just previously dipped in boiling water. It is important to seal the jars before the temperature has dropped appreciably.

To make sure that the fruit will keep, the bottles should then be immersed in a pan of hot water (fitted with a false bottom), the water brought to the boil and boiled for 5 minutes.

Fruit purée. The prepared fruit is cooked and sieved, then re-boiled before bottling by this method. Jars of tomato purée should be kept in the boiling water for 10 minutes.

Purée may be made in a liquidiser or food processor, but will still need sieving to get rid of pips and skins.

• Testing the seal

When the bottles are quite cold, preferably the day after processing, the clips or screw-bands should be removed and each jar carefully lifted by the lid. If the lid remains firm, a vacuum has formed, but if the lid comes off, the jar has not sealed properly and the fruit will not keep. If you wish to re-seal, look for the fault that has caused the failure to form a vacuum, remedy this, and re-process as soon as possible. Remember that even a sealed jar may not keep unless the processing has been done correctly; that is why it is important to follow the instructions carefully.

• Storing bottled fruit

It is often convenient to store the screw-bands on the bottles. These should be washed, dried thoroughly, rubbed on the inside with a little oil, and put on loosely. Spring clips should never be stored on the bottles.

Fruit stored in a warm place will lose its colour, so that the jars should be kept in a cool place away from sunlight. Labelling the jars with the date of bottling will help to ensure their being used in rotation.

• Vegetables

Vegetables should never be bottled in the same way as fruit. Much higher temperatures of processing are required, otherwise there is a danger of poisonous bacteria developing during storage, so that it is an impracticable process to undertake at home.

• USING BOTTLED FRUIT AND PURÉE •

Bottled fruit can be used in the same way as fresh and frozen fruit, but it is already cooked and should not be subjected to long cooking processes. To use bottled fruit cold, just turn it out into a bowl and add sweetening if necessary before serving with cream, ice cream or custard. For a fruit flan, arrange drained fruit in a baked pastry flan case or sponge case. Heat the liquid from the bottle and thicken with a little arrowroot. Cool slightly before pouring over the fruit.

For fruit trifles, or for filling gâteaux, drain the fruit very well before adding to the dish (and use the drained liquid for flavouring fresh fruit when you cook it). For crumbles, drain the fruit and tip it into an ovenware dish, before covering with the crumble mixture and baking at 200°C/400°F/gas mark 6 until the topping is crisp and golden.

Purée may be just turned out of the bottle and mixed with whipped cream or custard to make a fruit fool, or it can be used as the base of a fruit ice cream. Bottled fruit purée may also be used as sauce for vanilla ice cream, served with milk puddings or steamed puddings.

• PROCESSING TIMES FOR BOTTLED FRUIT •

Process	SLOW WATER BATH I		QUICK WATER BATH II
Method	Raise from cold to temperature in 90 minutes. Maintain as below.		Raise from warm (50°C/122°F) to simmer (95°C/203°F) in 25–30 minutes. Maintain as below.
Syrup or water	Cold added before processing.		Hot (70°C/155°F) before processing.
FRUIT	*Temperature*	*Maintain*	*Maintain*
Soft fruit, normal packs	85°C/185°F	10 minutes	2 minutes
Gooseberries (for pies)	85°C/185°F	10 minutes	2 minutes
Rhubarb (for pies)	85°C/185°F	10 minutes	2 minutes
Apple slices	85°C/185°F	10 minutes	2 minutes
Soft fruit, tight packs except Strawberries)	90°C/194°F	15 minutes	10 minutes
Gooseberries (for dessert)	90°C/194°F	15 minutes	10 minutes
Rhubarb (for dessert)	90°C/194°F	15 minutes	10 minutes
Stone fruit, dark, whole	90°C/194°F	15 minutes	10 minutes
Stone fruit, light whole	90°C/194°F	15 minutes	10 minutes
Citrus fruit, normal packs	90°C/194°F	15 minutes	10 minutes
Apples, solid packs	90°C/194°F	15 minutes	20 minutes
Citrus fruit, tight packs	90°C/194°F	15 minutes	20 minutes
Nectarines	90°C/194°F	15 minutes	20 minutes
Peaches	90°C/194°F	15 minutes	20 minutes
Pineapple	90°C/194°F	15 minutes	20 minutes
Plums, halved	90°C/194°F	15 minutes	20 minutes
Strawberries, soaked	90°C/194°F	15 minutes	20 minutes
Pears	95°C/203°F	30 minutes	40 minutes
Tomatoes, whole	95°C/203°F	30 minutes	40 minutes
Tomatoes, solid pack	95°C/203°F	40 minutes	50 minutes

PRESSURE COOKED III	SLOW OVEN IV		MODERATE OVEN V	
Raised from hot to 5lb/2.5kg (low) pressure in 5–10 minutes. Maintain as below. Cool 10 minutes before opening.	Pre-heat 15 minutes to 130°C/250°F. Process time varies with quantity in oven as below.		Pre-heat 15 minutes to 150°C/300°F. Process time varies with quantity in oven as below.	
Boiling added before processing	Boiling added after processing		Boiling added before processing	
Maintain	Quantity processed		Quantity processed	
	500g–2kg/1–4lb	2½–5kg/5–10lb	500g–2kg/1–4lb	2½–5kg/5–10lb
1 minute	44–55 minutes	60–75 minutes	30–40 minutes	45–60 minutes
1 minute	45–55 minutes	60–75 minutes	30–40 minutes	45–60 minutes
1 minute	Not recommended		30–40 minutes	45–60 minutes
1 minute	Not recommended		30–40 minutes	45–60 minutes
1 minute	55–70 minutes	75–90 minutes	40–50 minutes	55–70 minutes
1 minute	55–70 minutes	75–90 minutes	40–50 minutes	55–70 minutes
1 minute	Not recommended		40–50 minutes	55–70 minutes
1 minute	Not recommended		40–50 minutes	55–70 minutes
10 minutes	Not recommended		40–50 minutes	55–70 minutes
10 minutes	Not recommended		40–50 minutes	55–70 minutes
3–4 minutes	Not recommended		50–60 minutes	65–80 minutes
3–4 minutes	Not recommended		50–60 minutes	65–80 minutes
3–4 minutes	Not recommended		50–60 minutes	65–80 minutes
3–4 minutes	Not recommended		50–60 minutes	65–80 minutes
3–4 minutes	Not recommended		50–60 minutes	65–80 minutes
3–4 minutes	Not recommended		50–60 minutes	65–80 minutes
3–4 minutes	Not recommended		50–60 minutes	65–80 minutes
5 minutes	Not recommended		60–70 minutes	75–80 minutes
5 minutes	80–100 minutes	105–125 minutes	60–70 minutes	75–90 minutes
15 minutes	Not recommended		70–80 minutes	85–100 minutes

Allow extra time for large jars processed by Methods I and II as follows:
1½–2kg/3 and 4lb increase by 5 minutes all packs except tomatoes, solid pack increase by 10 minutes.
2½–3kg/5 and 6lb increase by 10 minutes all packs except tomatoes, solid pack increase by 20 minutes.
3½–4kg/7 and 8lb increase by 15 minutes all packs except tomatoes, solid pack increase by 30 minutes.

FREEZING

Freezing is probably the most popular method of home preserving, as preparation methods are easily understood and followed. Most foods, including cooked dishes, can be frozen and will retain their original flavour and texture.

• BASIC FREEZING RULES •

1 All food must be prepared without delay. Meat, poultry and fish should be prepared as soon as they are ready, and in cool conditions. Fruit and vegetables suffer if they are not processed immediately after picking. Cooked dishes should be cooled rapidly for freezing, and leftovers processed as soon as the meal is over.
2 Food must be packed in moisture-vapour-proof wrapping and firmly sealed. Air should be removed, but a headspace must be left in rigid containers for the expansion of liquids. Strongly-smelling or highly-flavoured foods should be overwrapped.
3 Food must be frozen rapidly. Animal products and cooked foods in particular can deteriorate quickly. Fast freezing slows up the formation of ice crystals which can spoil the texture of food.
4 Only the amount of fresh food which can be frozen safely within 24 hours, should be frozen at one time. This is usually the amount which would normally fill one-tenth of the total storage capacity. Overloading the freezer with fresh food will result in slower freezing.
5 Food must be stored at the correct temperature of −18°C/0°F. High quality storage life is only ensured when food is stored at this temperature.
6 Food must be thawed and/or cooked correctly. Rapid thawing results in toughness and loss of texture and flavour.

• Foods unsuitable for freezing

Hard-boiled eggs (including Scotch eggs).
Soured cream and single cream (less than 40 per cent butterfat).
Custards (including tarts).
Soft meringue toppings.
Mayonnaise and salad dressings.
Milk puddings.

Royal icing and frostings without fat.
Salad vegetables with a high water content, e.g. lettuce.
Old boiled potatoes (potatoes can be frozen mashed or baked).
Whole eggs in shells.

◆ Packaging materials

Packaging must be easy to handle, must not be liable to split, burst or leak, and must withstand the low temperature at which the freezer is maintained. Choice may be made from wrappings and containers of foil, polythene and waxed materials.

Foil is one of the most useful wrappings for the freezer, and may be cleaned and used again. Food can often be cooked in the foil used for freezer storage. Wrap sheet foil around awkward parcels of food, and use it for packing meat, poultry, cakes and pies. Foil may also be used to make lids for containers. Foil pie and pudding dishes can be used for baking, freezing and reheating dishes, and deep foil containers with their own lids are useful for casseroles.

Rigid plastic boxes may be used for free-flow vegetables and fruit, and for liquid items. They stack neatly and can be used many times. The lids must be airtight and may be sealed with freezer tape for extra protection.

Polythene should be heavy-gauge for freezer use. Gusseted bags are easier to pack and are suitable for most foods. Bags can be sealed with heat or a twist fastening, and air must be excluded. Sheeting is useful for wrapping meat, poultry and large cakes or pies, but it must be sealed with freezer tape.

Waxed containers are useful for fruit and for liquids, but they stain and retain smells, and are not always easy to use a second time.

◆ Open-freezing

Fruit and vegetables may be open-frozen on a special open-freezing tray without covering; iced cakes or fragile pies may be frozen this way. Food should be packed when it is frozen solid. This means that fruit and vegetables will be free-flowing, and the surface of cooked dishes and cakes will not be spoiled.

◆ Excluding air

Whether food is open-frozen or packed before freezing, all air must be excluded from packages. Air may be pressed out of soft packages by hand, or may be extracted by sucking through a drinking straw, or with a special pump. Air pockets in cartons can be released by plunging a knife into the contents two or three times.

• Headspace

Liquids expand when frozen so a space must be left in containers which hold liquid foods. Allow a headspace of 12mm/½in in wide-topped containers and 18mm/¾in in narrow-topped containers. When packing fruit, allow 12mm/½in for all dry packs; 12mm–2.5cm/½–1in per 600ml/1 pint for wide-topped wet packs; 1.8–2.5cm/¾–1in per 600ml/1 pint for narrow-topped wet packs. Allow double headspace for 1 litre/2 pint containers.

• High quality storage life

While frozen food is stored at −18°C/0°F slow changes are taking place in colour, texture and flavour. 'High quality storage life' is judged as the time taken for these changes to be noticeable in comparison with the equivalent fresh food. Opinions vary as to the length of this storage life, in relation to a person's senses of taste and smell. Foods with a high fat content or with highly volatile ingredients have a shorter life than vegetables, for instance, and should therefore only be stored for a relatively short time.

• VEGETABLES •

All vegetables to be frozen should be young and tender. Shop-bought vegetables are generally too old to be worth freezing, but a few seasonal delicacies such as aubergines and peppers are worth the trouble.

• Blanching

Vegetables to be frozen must be blanched. Blanching at high heat stops the enzymes from affecting quality, flavour, colour and nutritive value during storage.

All vegetables must first be washed thoroughly in cold water, then cut or sorted into similar sizes. If more are picked than can be dealt with, they should be put into polythene bags in a refrigerator until they can be processed.

Blanch 500g/1lb vegetables at a time to ensure thoroughness and to prevent a quick change in the water temperature. Use a saucepan holding at least 4.5 litres/8 pints water. Completely immerse the vegetables in the saucepan of fast-boiling water; cover tightly and keep the heat high under the saucepan until blanching is completed. Check carefully the time needed for each vegetable and time the blanching from when water returns to boiling point. As soon as the full blanching time has elapsed, remove vegetables and drain at once.

Cooling must be done immediately after blanching, and must be very

Bottled Rhubarb (pages 7–8).

SEALING DISC
FOR
KILNER
Dual Purpose
JAR
KILNER
KILNER
Dual Purpose

thorough. The time taken is generally equal to the blanching time if a large quantity of cold water is used. Vegetables which are not cooled quickly become mushy. After cooling in the water, the vegetables should be thoroughly drained. Pack the cooled vegetables in usable quantities in bags or boxes.

Artichokes (Globe). (a) Remove outer leaves. Wash, trim stalks and remove 'chokes'. Blanch in 4·5 litres/8 pints water with 15ml/1tbsp lemon juice for 7 minutes. Cool and drain upside down. Pack in boxes. (b) Remove all green leaves and 'chokes'. Blanch artichoke hearts for 5 minutes. High quality storage life: 12 months.

Asparagus. Wash and remove woody portions and scales. Grade for size and cut in 15cm/6in lengths. Blanch 2 minutes (small spears); 3 minutes (medium spears); 4 minutes (large spears). Cool and drain. Pack in boxes. High quality storage life: 9 months.

Aubergines. Use mature, tender, medium-sized. (a) Peel and cut in 2.5cm/1in slices. Blanch 4 minutes, chill and drain. Pack in layers separated by paper in boxes. (b) Coat slices in thin batter, or egg and breadcrumbs. Deep-fry, drain and cool. Pack in layers in boxes. High quality storage life: (a) 12 months, (b) 1 month.

Beans (Broad). Use small young beans. Shell and blanch for 1½ minutes. Pack in bags or boxes. High quality storage life: 12 months.

Beans (Dwarf). Use young tender beans. Remove tops and tails and use whole or cut into 2·5cm/1in pieces. Blanch whole beans 3 minutes, cut beans 2 minutes. Cool and pack. High quality storage life: 12 months.

Beans (Runner). Do not shred the beans, but cut them in pieces and blanch them for 2 minutes. Cool and pack. High quality storage life: 12 months.

Beetroot. Use very young beetroot, under 7.5cm/3in across. They must be completely cooked in boiling water until tender. Rub off skins and pack in boxes, either whole or cut in slices or diced. High quality storage life: 6 months.

Broccoli. Use green, compact heads with tender stalks 2.5cm/1in thick or less. Trim stalks and remove outer leaves. Wash well and soak in salt water for 30 minutes (2 tsp salt to 4.5 litres/8 pints water). Wash in fresh water, and cut into sprigs. Blanch 3 minutes (thin stems); 4 minutes (medium stems); 5 minutes (thick stems). Pack into boxes or bags, alternating heads. High quality storage life: 12 months.

Brussels sprouts. Grade small compact heads. Clean and wash well. Blanch 3 minutes (small); 4 minutes (medium). Cool and pack in bags or boxes. High quality storage life: 12 months.

Salted Beans (page 43).

Cabbage (Green and Red). Use crisp young cabbage. Wash and shred finely. Blanch 1½ minutes. Pack in bags. If preferred, red cabbage may be cooked with vinegar and spices and frozen as a complete dish. High quality storage life: 6 months.

Carrots. Use very young carrots. Wash and scrape. Blanch 3 minutes for small whole carrots, sliced or diced carrots. Pack in bags or boxes. High quality storage life: 12 months.

Cauliflower. Use firm compact heads with close white flowers. Wash and break into sprigs. Blanch 3 minutes in 4.5 litres/8 pints water with 15ml/1 tbsp lemon juice. High quality storage life: 6 months.

Celery. (a) Use crisp young stalks. Scrub well and remove strings. Cut in 2.5cm/1in lengths and blanch 2 minutes. Cool and drain and pack in bags. (b) Prepare as above, but pack in boxes with water used for blanching, leaving 1.25cm/½in headspace. High quality storage life: 6 months.

Chicory. Wash well and remove outer leaves. Blanch 3 minutes and cool in cooking liquid. Pack in the blanching liquid in boxes, leaving 1.25cm/½in headspace. High quality storage life: 6 months.

Corn-on-the-cob. (a) Use fresh tender corn. Remove leaves and threads and grade cobs for size. Blanch 4 minutes (small cobs); 6 minutes (medium cobs); 8 minutes (large cobs). Cool and dry. Pack individually in foil or freezer paper. Freeze and pack in bags for storage. (b) Blanch cobs and scrape off kernels. Pack in boxes, leaving 1.25cm/½in headspace. High quality storage life: 12 months.

Courgettes see *Marrow*.

Fennel. Scrub well and remove any dirt under running water. Blanch 3 minutes, retaining blanching liquid for packing. Pack in rigid containers with liquid, leaving 1.25cm/½in headspace. High quality storage life: 6 months.

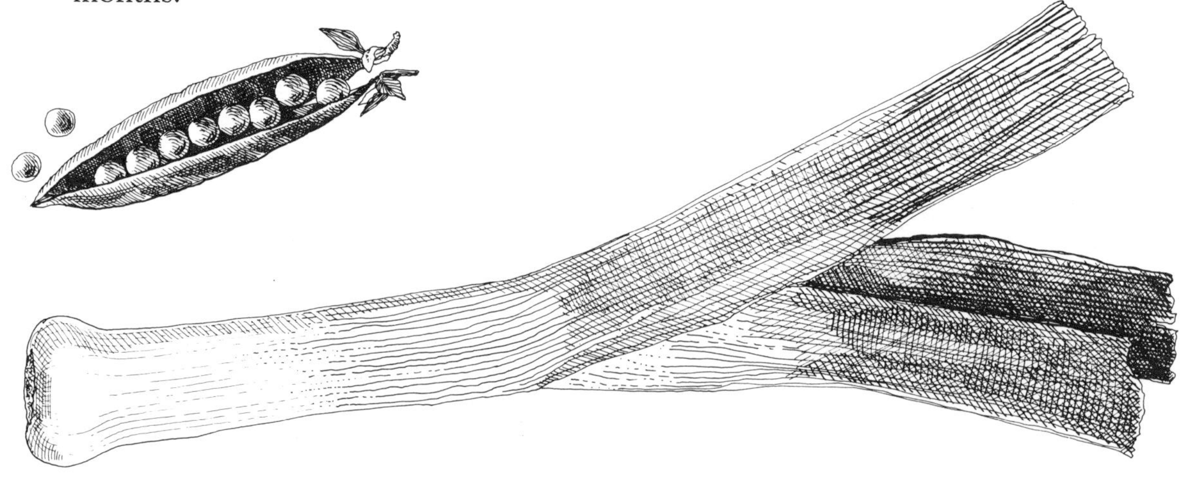

Leeks. Trim off roots and green stems. Wash very well and remove dirty outer layers. Cut either finely or coarsely into even lengths. Blanch finely cut leeks for 1½ minutes; coarsely cut leeks for 3 minutes. Cool thoroughly and drain, or pack in blanching liquid. High quality storage life: 12 months.

Marrow. (a) Cut young marrows or courgettes in 1.25cm/½in slices without peeling. Blanch 3 minutes and pack in boxes, leaving 1.25cm/½in headspace. (b) Peel and seed large marrows. Cook until soft, mash and pack in boxes. High quality storage life: 6 months.

Mushrooms. (a) Wipe but do not peel. Stalks may be frozen separately. Pack in bags. (b) Grade and cook in butter for 5 minutes. Allow 75g/3oz butter to 500g/1lb mushrooms. Cool quickly, take off excess fat, and pack in boxes. High quality storage life: (a) 3 months, (b) 2 months.

Onions. (a) Peel, chop and pack in small boxes. Overwrap. (b) Cut in slices and wrap in foil or freezer paper, dividing layers with paper. Overwrap. (c) Chop or slice, blanch 2 minutes. Cool and drain. Pack in boxes. Overwrap. (d) Leave tiny onions whole. Blanch 4 minutes. Pack in boxes. Overwrap. High quality storage life: 2 months.

Parsnips. Use young parsnips. Trim and peel. Cut into narrow strips or dice. Blanch 2 minutes. Pack in bags or boxes. High quality storage life: 12 months.

Peas, Green. Use young sweet peas. Shell. Blanch 1 minute, shaking basket to distribute heat. Cool and drain. Pack in boxes or bags. High quality storage life: 12 months.

Peas (Edible Pods). Use flat tender pods. Wash well. Remove ends and strings. Blanch 30 seconds in small quantities. High quality storage life: 12 months.

Peppers (Green and Red). Wash well. Cut off stems and caps, and remove seeds and membranes. Blanch 2 minutes (slices); 3 minutes (halves). Pack in boxes or bags. High quality storage life: 12 months.

Potatoes. (a) Scrape and wash new potatoes. Blanch 4 minutes. Cool and pack in bags. (b) Slightly undercook new potatoes. Drain, toss in butter, cool and pack in bags. (c) Mash potatoes with butter and hot milk. Pack in boxes or bags. (d) Form potatoes into croquettes or Duchesse Potatoes. Cook, cool and pack in boxes. (e) Fry chips in clean fat for 4 minutes. Do not brown. Cool and pack in bags. High quality storage life: (a) 12 months, (b)–(e) 3 months.

Spinach. Use young tender spinach. Remove stems. Wash very well. Blanch 2 minutes, shaking basket so the leaves separate. Cool and press out moisture. Pack in boxes or bags. High quality storage life: 12 months.

Tomatoes. (a) Wipe tomatoes and remove stems. Grade and pack in small quantities in bags. (b) Skin and core tomatoes. Simmer in own juice for 5 minutes until soft. Sieve, cool and pack in boxes. (c) Core tomatoes and cut in quarters. Simmer with lid on for 10 minutes. Put through muslin. Cool juice and pack in boxes, leaving 2.5cm/1in headspace. High quality storage life: 12 months.

Turnips. (a) Use small, young, mild turnips. Peel and dice. Blanch 2½ minutes. Cool and pack in boxes. (b) Cook turnips until tender. Drain and mash. Pack in boxes, leaving 1.25cm/½in headspace. High quality storage life: (a) 12 months, (b) 3 months.

Vegetable purée. Cook and sieve vegetables. Pack in boxes. Small quantities can be frozen in ice-cube trays, and the cubes transferred to bags for easy storage. High quality storage life: 3 months.

• FRUIT •

The best results are obtained from fully flavoured fruits, particularly berries. In general, fruit for freezing should be of top quality; over-ripe fruit will be mushy; unripe fruit will be tasteless and poorly coloured.

Home-produced fruit should be frozen on the same day as picking, while fruit from shop or market should only be bought in manageable quantities which can be handled in a short space of time.

• Unsweetened dry pack

This pack can be used for fruit to be used in pies, puddings and jams, or for people on a sugar-free diet. It should not be used for fruit which discolours badly during preparation, as sugar helps to retard the action of the enzymes which cause darkening.

To pack fruit by this method, wash and drain and pack into boxes or bags. Do not use excess water in cleaning the fruit. Seal and freeze and label carefully. For an unsweetened pack, it is better to open-freeze fruit. Spread it out on metal or plastic trays for fast freezing. When frozen, pour into polythene bags or boxes for storage.

• Dry sugar pack

This is a good method for crushed or sliced fruit, or for soft juicy fruit. The fruit should be washed and drained and may be packed by two methods: (a) mix fruit and sugar in a bowl with a silver spoon, adjusting sweetening to tartness of fruit (average 1½kg/3lb fruit to 500g/1lb sugar). Pack fruit into containers, leaving 1.25cm/½in headspace, seal and freeze, and label carefully; (b) pack fruit in layers using the same proportion of fruit and sugar; start with a layer of fruit, sprinkle with sugar, then more fruit and sugar, leaving 1.25cm/½in headspace. Seal and freeze, labelling carefully.

◆ Syrup pack

This method is best for non-juicy fruits and those which discolour easily. Syrup is normally made from white sugar and water. (Honey may be used, but it flavours the fruit strongly. Brown sugar may be used, but affects the colour of the fruit.) A medium syrup or 40 per cent syrup (325g/11oz sugar to 600ml/1 pint water) is best for most purposes as a heavier syrup tends to make the fruit flabby. The sugar must be completely dissolved in boiling water, then cooled, and is best stored in a refrigerator for a day before use. The fruit should be packed into containers and covered with syrup, leaving 1.25–2.5cm/½–1in headspace. To prevent discoloration of the fruit, a piece of greaseproof paper or freezer film should be pressed down over the fruit into the syrup before sealing, freezing and labelling.

◆ Discoloration

Apples, peaches and pears are particularly subject to discoloration during preparation, storage and thawing. In general, fruit which has a lot of vitamin C darkens less easily, so adding lemon juice or citric acid to the sugar pack will help to arrest darkening. Use the juice of 1 lemon to 900ml/1½ pints water or 5ml/1 tsp citric acid to each 500g/1lb sugar in dry pack. Fruit purée in particular is subject to darkening. For the same reason, fruit should be eaten immediately on thawing, or while a few ice crystals remain.

◆ Jam fruit

Any fruit can be packed for use in jam-making later. Pack without sweetening, and allow 10 per cent extra fruit in the recipe when making the jam, as there is a slight pectin loss in frozen fruit.

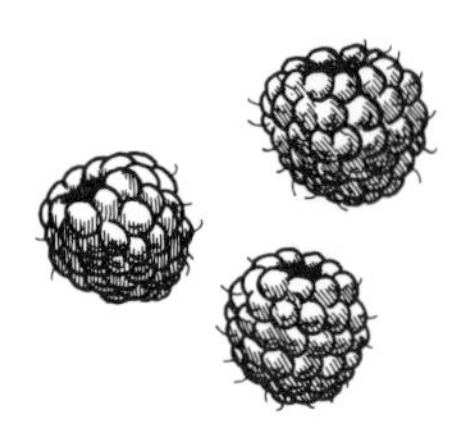

◆ Fruit purée

The fruit should not be over-ripe or bruised. Raw fruit such as raspberries or strawberries should be sieved, to remove all the pips. Other fruit can be put in a covered dish in the oven to start the juice running before the fruit is sieved. Purée can be made from cooked fruit but must be well cooled before freezing. This purée should be sweetened, as if for immediate use.

◆ Fruit juices

Ripe fruit can be turned into juice, and frozen. Citrus fruit juice can also be frozen. Non-citrus fruit should be carefully checked for any bruising or insects, then mashed with a silver fork. For every 4 cups of fruit, allow 1 cup of water and simmer gently for 10 minutes. Strain through a jelly bag or cloth, and cool completely before freezing. Juices can be frozen unsweetened or sweetened. Freeze in a rigid container, leaving 1.25cm/½in headspace, or in ice-cube trays, wrapping each cube in foil and storing in quantities in polythene bags. Apple juice can be made, using 300ml/½ pint water to each 1kg/2lb apples, by simmering leftover peelings in water; it should not be sweetened before freezing since fermentation sets in quickly.

Citrus fruit juices can easily be prepared from good-quality fruit which is heavy in the hand for its size. Freeze in rigid containers, leaving 2·5cm/1in headspace.

◆ Thawing frozen fruit

Unsweetened fruit packs take longer to thaw than sweetened ones; fruit in dry sugar thaws most quickly of all. All fruit should be thawed in its container unopened; and all fruit is at its best when just thawed. Fruit to use with ice cream should be only partly defrosted. To cook frozen fruit, thaw until pieces can just be separated and put into a pie; if fruit is to be cooked in a saucepan, it can be put into the pan in its frozen state. For every 500g/1lb fruit packed in syrup, allow 6–8 hours thawing time in the refrigerator, 2–4 hours thawing at room temperature, or ½–1 hour if the pack is placed in a bowl of cold water. For defrosting by microwave, follow manufacturers' instructions.

Apples. Peel, core and drop in cold water. Cut into twelfths or sixteenths. Pack in bags or boxes. (a) Dry sugar pack (250g/8oz sugar to 1kg/2lb fruit). (b) 40 per cent syrup pack. (c) Sweetened cooked purée. High quality storage life: (a) 8 months, (b) 8 months, (c) 4 months.

Apricots. (a) Peeled and halved in dry sugar pack (125g/4oz sugar to 500g/1lb fruit) or 40 per cent syrup pack. (b) Peeled and sliced in 40 per cent syrup pack. (c) Sweetened purée (very ripe fruit). High quality storage life: (a) 12 months, (b) 12 months, (c) 4 months.

Bananas. Mash with sugar and lemon juice (250g/8oz sugar to 45ml/3 tbsp lemon juice to 3 breakfastcups banana pulp). Pack in small containers. High quality storage life: 2 months.

Blackberries. Wash dark glossy ripe berries and dry well. (a) Fast-freeze unsweetened berries on trays and pack in bags. (b) Dry sugar pack (250g/8oz sugar to 1kg/2lb fruit). (c) Sweetened purée (raw or cooked fruit). High quality storage life: 12 months.

Cherries. Put in chilled water for 1 hour; remove stones. Pack in glass or plastic containers, as cherry juice remains liquid and leaks through waxed containers. (a) Dry sugar pack (250g/8oz sugar to 1kg/2lb stoned cherries). (b) 40 per cent syrup pack for sweet cherries. (c) 50 per cent or 60 per cent syrup pack for sour cherries. High quality storage life: 12 months.

Currants, Black, Red and White. Prepare black, red or white currants by the same methods. Strip fruit from stems with a fork, wash in chilled water and dry gently. Currants can be fast-frozen on trays and the stalks stripped off before packing. This makes the job easier. (a) Dry unsweetened pack. (b) Dry sugar pack (250g/8oz sugar to 500g/1lb currants). (c) 40 per cent syrup pack. (d) Sweetened cooked purée (particularly blackcurrants). High quality storage life: (a) 12 months, (b) 12 months, (c) 12 months, (d) 4 months.

Gooseberries. Wash in chilled water and dry. For pies, freeze fully ripe fruit; for jam, fruit may be slightly under-ripe. (a) Dry unsweetened pack. (b) 40 per cent syrup pack. (c) Sweetened purée. High quality storage life: (a) 12 months, (b) 12 months, (c) 4 months.

Grapefruit. Peel; remove pith; cut into segments. (a) Dry sugar pack (250g/8oz sugar to 2 breakfastcups segments). (b) 50 per cent syrup pack (500g/1lb sugar to 600ml/1 pint water). High quality storage life: 12 months.

Grapes. Pack seedless varieties whole. Skin and seed other types. Pack in 30 per cent syrup pack (175g/7oz sugar to 600ml/1 pint water). High quality storage life: 12 months.

Kumquats. Wrap whole fruit in foil or use 50 per cent syrup pack. Use immediately after thawing. High quality storage life: (a) 2 months if unsweetened, (b) 12 months in syrup.

Lemons and Limes. Peel the fruit, cut it in slices, and pack in 20 per cent syrup pack (125g/4oz sugar to 600ml/1 pint water). High quality storage life: 12 months.

Loganberries. Wash berries and dry well. (a) Fast-freeze unsweetened berries on trays and pack in bags. (b) Dry sugar pack (250g/8oz sugar to 1kg/2lb fruit). (c) 50 per cent syrup pack. (d) Sweetened purée (cooked fruit). High quality storage life: 12 months.

Mangoes. Peel and slice ripe fruit. Pack in 50 per cent syrup, allowing 1 tbsp lemon juice to 600ml/1 pint syrup. High quality storage life: 12 months.

Melons. Cut into cubes or balls. Toss in lemon juice and pack in 30 per cent syrup. High quality storage life: 12 months.

Oranges. Seville oranges may be frozen whole in their skins in polythene bags for marmalade. Peel and divide into sections or cut into slices. (a) Dry sugar pack (250g/8oz sugar to 3 breakfastcups sections or slices). (b) 30 per cent syrup pack. (c) Pack slices in slightly sweetened fresh orange juice. High quality storage life: 12 months.

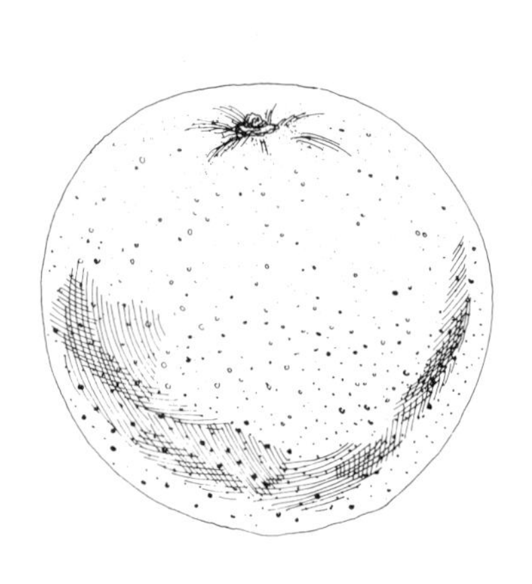

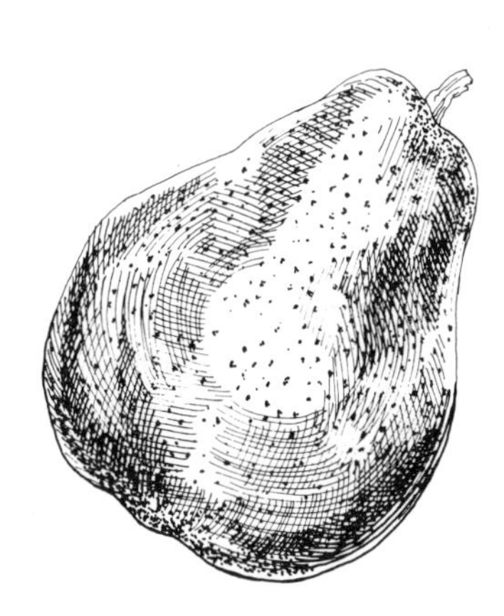

Peaches. Work quickly as fruit discolours. Peel, cut in halves or slices and brush with lemon juice. (a) 40 per cent syrup back. (b) Sweetened purée (fresh fruit) with 15ml/1 tbsp lemon juice to 500g/1lb fruit. High quality storage life: 12 months.

Pears. Pears should be ripe, but not over-ripe. They discolour quickly and do not retain their delicate flavour in the freezer. Peel and quarter fruit, remove cores, and dip pieces in lemon juice. Poach in 30 per cent syrup for 1½ minutes. Drain and cool. Pack in cold 30 per cent syrup. High quality storage life: 12 months.

Pineapple. Use ripe fruit. Peel and cut into slices or chunks. (a) Dry pack unsweetened. (b) Dry sugar pack (125g/4oz sugar to 500g/1lb fruit). (c) 30 per cent syrup. High quality storage life: 12 months.

Plums. Wash in chilled water and dry. Cut in half and remove stones. Pack in 40 per cent syrup. High quality storage life: 12 months.

Raspberries. (a) Dry unsweetened pack. (b) Dry sugar pack (125g/4oz sugar to 500g/1lb fruit). (c) 30 per cent syrup pack. (d) Sweetened purée (fresh fruit). High quality storage life: 12 months.

Rhubarb. Wash sticks in cold running water, and trim to required length. (a) Blanch sticks 1 minute, then wrap in foil or polythene. (b) 40 per cent syrup pack. (c) Sweetened cooked purée. High quality storage life: 12 months.

Strawberries. Use ripe, mature and firm fruit. Pick over fruit, removing hulls. (a) Grade for size in dry unsweetened pack. (b) Dry sugar pack (125g/4oz sugar to 500g/1lb fruit). Fruit may be sliced or lightly crushed. (c) 40 per cent syrup pack for whole or sliced fruit. (d) Sweetened purée (fresh fruit). High quality storage life: 12 months.

• MEAT •

Many authorities feel that fresh meat should not be frozen in domestic freezers, since it is not possible to achieve the very low temperatures thought necessary for successful freezing. This point should be thought about carefully when buying in bulk for the freezer.

Meat must be of good quality whatever the cut, and must be properly hung (beef 8–12 days; lamb 5–7 days; pork and veal chilled only).

• Preparation for freezing meat

Bulk supplies of meat should be packaged in quantities which can be used up on a single occasion if possible. Ideally, meat should be boned and the surplus fat removed. If the bones are not removed, the ends should be wrapped in several layers of greaseproof paper to avoid piercing freezer wrappings. Meat must be carefully labelled for identification. Air must be excluded from the packages so that the freezer wrap can touch the surface of the meat all over.

If a whole animal or a variety of different meats are being prepared for freezing at one time, the offal should be processed first, then pork, veal and lamb, and finally beef as this will keep best under refrigeration if delays occur.

The wrapping for meat must be strong, since oxygen from the air which may penetrate wrappings affects fat and may cause rancidity (pork is the most subject to this problem). In addition to moisture-vapour-proof wrapping, an overwrap of brown paper, greaseproof paper or stockinette will protect packages and will guard against punctures. Place the label on the outside of this wrapping.

Ham. Package uncooked ham in the piece rather than sliced. Pack in freezer paper, foil or polythene, and overwrap. Vacuum-packed joints may be frozen in packing. Storage life is limited as salt causes rancidity. High quality storage life: 3 months (whole); 1 month (sliced).

Hearts, Kidneys, Sweetbreads, Tongue. Wash and dry offal thoroughly. Remove blood vessels and pipes. Wrap in cellophane or polythene and pack in bags or boxes. Off-flavours may develop if offal is not packed with care. Tongue may be cooked and frozen whole if preferred. High quality storage life: 2 months.

Joints. Trim surplus fat. Bone and roll if possible. Pad sharp bones. Wipe meat. Pack in polythene bag or sheeting, freezer paper or foil. Remove air. Freeze quickly. High quality storage life: beef 12 months; lamb 9 months; pork 6 months; veal 9 months.

Liver. Package whole or in slices. Separate slices with greaseproof paper or freezer film. High quality storage life: 2 months.

Minced meat. (a) Use good quality mince without fat. Pack in bags or boxes. Do not add salt. Remove air. Freeze quickly. (b) Shape mince into patties, separated, and pack in bags or boxes. Remove air. Freeze quickly. High quality storage life: 2 months.

Sausages and sausage-meat. Omit salt in preparation. Pack in usable quantities. Wrap tightly in freezer paper, foil or polythene. High quality storage life: 1 month.

Steaks and chops. Package in usable quantities. Separate pieces of meat with greaseproof paper or freezer film. Pack in a polythene bag or sheeting, freezer paper or foil. Remove air. Freeze quickly. High quality storage life: 6–12 months (according to type of meat).

Tripe. Cut in 2.5cm/1in squares and pack tightly in bags or boxes. High quality storage life: 2 months.

Thawing meat. Thaw slowly in refrigerator. For defrosting by microwave, follow manufacturers' instuctions.

◆ BACON ◆

◆ Buying bacon for freezing

Freshness of bacon is the first vital step to successful freezing. Try and get the bacon on the day that the retailer gets delivery from his supplier. Determine storage period in relation to freshness, and reduce the recommended period if in doubt. Smoked bacon can be stored for longer than unsmoked bacon. The quicker bacon is frozen right through, the better it will be. It is therefore inadvisable to freeze pieces weighing more than 2.5kg/5lb.

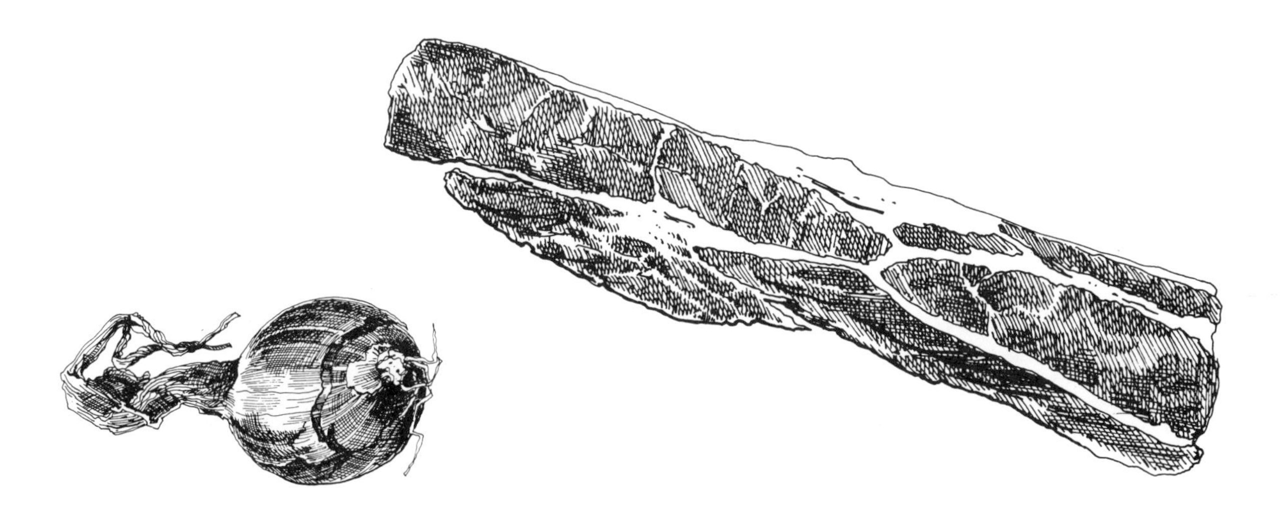

• Vacuum-packed bacon

Vacuum-packing of bacon is the ideal preparation for storage in the freezer because air has already been withdrawn from the packet.

A lot of vacuum-packed bacon is so marked; but as vacuum-packing can be confused with other types of wrapping, it is advisable to check this if one is in doubt. To prepare these packets for the freezer, inspect each one to ensure the vacuum is not damaged, i.e. the bacon should not be loose in the packet. Wrap the packets in foil, and label.

• Thawing and cooking frozen bacon

(a) *Joints.* Allow bacon plenty of time to thaw slowly, preferably in a refrigerator. Bacon can be thawed at room temperature before cooking. Time required depends on the thickness of the piece and the temperature. The wrapping should be removed as soon as possible during thawing. If time is short, joints may be thawed in running water, but should be wrapped in a plastic bag to prevent them getting wet. For defrosting by microwave, follow manufacturers' instructions.

(b) *Bacon rashers and small pieces* may be thawed overnight in the refrigerator or at room temperature. Cook immediately following instructions on the packet. All frozen bacon should be cooked immediately it has thawed. The usual cooking methods – boiling, grilling, frying and baking are suitable. Once cooked, the bacon will keep 1–2 days in a refrigerator.

High quality storage life: 1. Bacon joints wrapped as recommended: (a) Smoked bacon up to 8 weeks, (b) Unsmoked bacon up to 5 weeks; 2. Vacuum-packed bacon joints up to 10 weeks; 3. Vacuum-packed rashers or steaks up to 10 weeks; 4. Foil-wrapped rashers, chops or steaks, smoked, 2–4 weeks only.

• POULTRY AND GAME •

Birds to be frozen should be in perfect condition. When the bird is plucked, it is important to avoid skin damage; if scalding, beware of over-scalding which may increase the chance of freezer-burn (grey spots occurring during storage). The bird should be cooled in a refrigerator or cold larder for 12 hours, drawn and completely cleaned. With geese and ducks, it is particularly important to make sure that the oil glands are removed as these will cause tainting.

Giblets have a storage life of only 2 months, so unless a whole bird is to be used within that time, it is not advisable to pack them inside the bird. Giblets should be cleaned, washed, dried and chilled, then wrapped in moisture-vapour-proof paper or a bag, excluding air, and frozen in batches. Livers should be treated in the same way.

Joints should be divided by two layers of freezer film. Bones of young birds may turn brown in storage, but this does not affect flavour or quality.

◆ Thawing poultry

Uncooked poultry must thaw completely before cooking. Thawing in the refrigerator will allow slow, even thawing; thawing at room temperature will be twice as fast but the product will be much less satisfactory. A 2–2.5kg/4–5lb chicken will thaw overnight in a refrigerator and will take 6 hours at room temperature. A turkey weighing 4.5kg/9lb will take 36 hours; as much as 3 days should be allowed for a very large bird. A thawed bird can be stored for up to 24 hours in a refrigerator.

For defrosting by microwave, follow manufacturers' instructions.

Game. Young game is best for freezing. All game should be hung to its required state before freezing, as hanging after thawing will result in the flesh going bad.

All game should be kept cool between shooting and freezing; care should be taken to remove as much shot as possible, and to make sure the shot wounds are thoroughly clean. Birds should be bled as soon as shot, and then hung to individual taste. After plucking and drawing, the cavity should be thoroughly washed and drained and the body wiped with a damp cloth. The birds should then be packed, cooled and frozen like poultry.

Chicken. Hang and cool. Pluck and draw and pack giblets separately. Truss whole bird or cut in joints. Chill 12 hours. Pack in bag, removing air. High quality storage life: 12 months.

Duck. Hang and cool. Remove oil glands. Pluck and draw and pack giblets separately. Chill 12 hours. Pack in bag, removing air. High quality storage life: 6 months.

Goose. Hang and cool. Remove oil glands. Pluck and draw and pack giblets separately. Chill 12 hours. Pack in bag, removing air. High quality storage life: 6 months.

Grouse (12 August–10 December); Partridge (1 September–1 February); Pheasant (1 October–1 February). Remove shot and clean wounds. Bleed as soon as shot, keep cool and hang to taste. Pluck, draw and truss. Pad bones. Pack in bag, removing air. High quality storage life: 6 months.

Hares and Rabbits. Clean shot wounds. Behead and bleed as soon as possible, collecting hare's blood if needed for cooking. Hang for 24 hours in a cool place. Skin, clean and wipe. Cut into joints and wrap each piece in cellophane. Pack in usable quantities in bags. Pack blood in box. High quality storage life: 6 months.

Livers. Clean, wash, dry and chill. Pack in bag, removing air. High quality storage life: 2 months.

Pigeons. Remove shot and clean wounds. Prepare and pack as feathered game. High quality storage life: 6 months.

Plover, Quail, Snipe, Woodcock. Remove shot and clean wounds. Prepare as other feathered game but do not draw. Pad bones. Pack in bag, removing air. High quality storage life: 6 months.

Turkey. Hang and cool. Pluck and draw. Truss whole or cut in joints. Chill for 12 hours. Pack in bag, removing air. High quality storage life: 6 months.

Venison. Clean shot wounds. Keep the carcass cold until butchered. Behead, bleed, skin and clean, wash and wipe flesh. Hang in a cool place for 5 days. Joint and pack in bags, removing air. Freeze the good joints, but prepare other cuts as cooked dishes for freezing. Thaw in wrappings in refrigerator for 4 hours. Remove from wrappings and put into marinade. Continue thawing, allowing 5 hours per 500g/1lb. High quality storage life: 12 months.

• FREEZING FISH AND SHELLFISH •

Only really fresh fish can be frozen, since it must be processed within 24 hours, so it is not advisable to freeze fish from a shop.

Since fish must be fresh, it must be cleaned as soon as it is caught. The fish should be killed at once, scaled if necessary and fins removed. Small fish can be left whole; large fish should have heads and tails removed, or can be divided into steaks. Flat fish and herrings are best gutted, and flat fish skinned and filleted. White fish should be washed well in salted water during cleaning, but fatty fish should be washed in fresh water.

Fish may be frozen in a number of ways:

• Dry pack

Separate pieces of fish with freezer film, pack in box or bag, seal and freeze. Be sure wrapping is in close contact with the fish to exclude air which will dry the fish and make it tasteless. Freeze quickly.

• Acid pack

Citric acid preserves the colour and flavour of fish; ascorbic acid is an anti-oxidant which stops the development of rancidity in fish. A chemist can provide an ascorbic-citric acid power, to be diluted in a proportion of 1 part powder to 100 parts of water. Dip fish into this solution, drain, wrap and seal.

• Solid ice pack

Small fish, steaks or fillets can be covered with water in refrigerator trays or loaf tins and frozen into solid blocks. The fish should be separated by double paper as usual. Remove ice blocks from pan, wrap in freezer paper and store. The fish can also be frozen in a solid ice pack in large waxed tubs; cover the fish completely to within 1.25cm/½in of container top and crumple a piece of freezer film over the top of the fish before closing the lid. The only advantage in this solid ice method is a saving of containers and wrapping material.

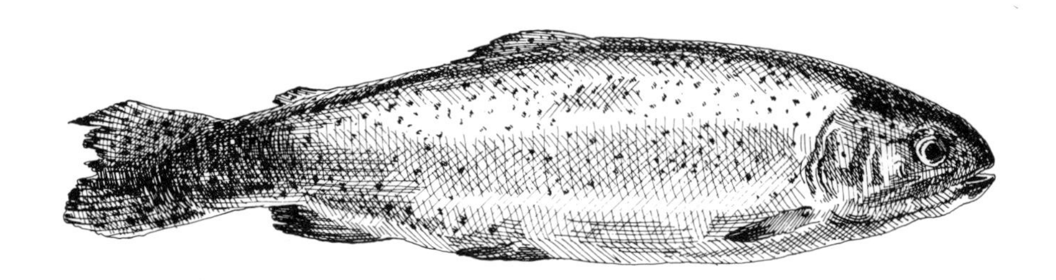

◆ Glazing a large fish

First clean the fish. Then place the unwrapped fish against the freezer wall in the coldest possible part of the freezer. When the fish is frozen solid, dip it very quickly into very cold water so a thin coating of ice will form. Return fish to freezer for an hour, and repeat process. Continue until ice has built up to 0·5cm/¼in thickness. The fish can be stored without wrappings for 2 weeks, but is better wrapped in polythene for longer storage. Sometimes a large whole fish may be wanted; if so, it can be frozen whole, but is best protected by 'glazing'. Salmon and salmon trout are obvious examples, or perhaps a haddock or halibut to serve stuffed for a party.

Crab. Cook, drain and cool. Clean crab and remove edible meat. Pack into boxes or bags. High quality storage life: 1 month.

Fatty fish (Haddock, Halibut, Mackerel, Salmon, Trout, Turbot). Clean. Fillet or cut in steaks if liked, or leave whole. Separate pieces of fish with double thickness of freezer film. Wrap in freezer paper, or put in box or bag. Be sure air is excluded, or fish will be dry and tasteless. Keep pack shallow. Freeze quickly. Large fish may be prepared in solid ice pack. High quality storage life: 1 month.

Lobster and Crayfish. Cook, cool and split. Remove flesh and pack into boxes or bags. High quality storage life: 1 month.

Mussels. Scrub very thoroughly and remove any fibrous matter sticking out from the shell. Put in a large saucepan and cover with a damp cloth. Put over medium heat about 3 minutes until they open. Cool in the pan. Remove from shells and pack in boxes, covering with their own juice. High quality storage life: 1 month.

Oysters. Open oysters and save liquid. Wash fish in salt water (5ml/1 tsp salt to 600ml/1 pint water). Pack in boxes, covering with own liquid. High quality storage life: 1 month.

Prawns. Cook and cool in cooking water. Remove shells. Pack tightly in boxes or bags. High quality storage life: 1 month.

Above: Smoked herring and haddock (page 51).
Below: Arbroath Smokies, a small cottage smoking industry.

Scallops. Open shells. Wash fish in salt water (5ml/1 tsp salt to 600ml/1 pint water). Pack in boxes covering with salt water, and leaving 1.25cm/½in headspace. High quality storage life: 1 month.

Shrimps. (a) Cook and cool in cooking water. Remove shells. Pack in boxes or bags. (b) Cook and shell shrimps. Pack in waxed boxes and cover with melted spiced butter. High quality storage life: 1 month.

Smoked fish (Haddock, Kippers, Mackerel, Salmon, Trout, etc.). Pack fish in layers with freezer film between. Keep pack shallow. High quality storage life: 2 months.

White fish (Cod, Plaice, Sole, Whiting). Clean. Fillet or cut in steaks if liked, or leave whole. Separate pieces of fish with freezer film. Wrap in polythene or put in box or bag. Be sure air is excluded, or fish will be dry and tasteless. Keep pack shallow. Freeze quickly. High quality storage life: 3 months.

• DAIRY PRODUCE •

Dairy produce should not be allowed to take up much freezer space. It can be useful however to freeze quantities of cheap fat or eggs; cheese left after large parties; leftover egg yolks or whites or cracked eggs or thick cream.

Butter or Margarine. Overwrap blocks in foil or polythene. High quality storage life: 6 months (unsalted); 3 months (salted).

Cheese. Freeze hard cheese in small portions (250g/8oz or less). Divide slices with freezer film and wrap in foil or polythene. Freeze grated cheese in polythene bags; the pieces remain separated. Freeze Camembert, Port Salut, Stilton, Danish Blue and Roquefort with careful sealing to avoid drying out and cross-contamination. Cut cheese while still half-frozen to prevent crumbling. High quality storage life: 3 months.

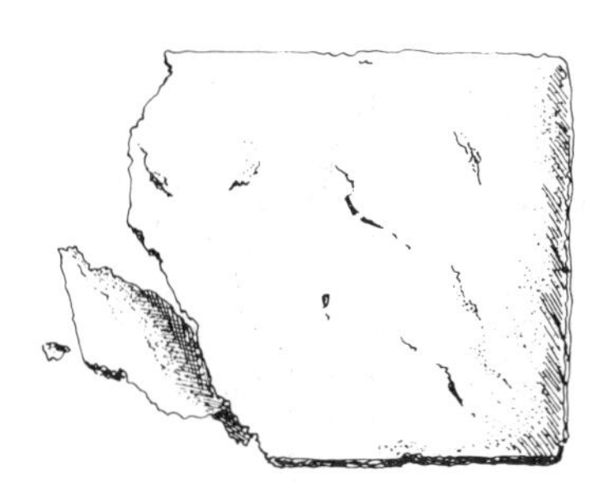

A selection of dried fruits (pages 69–70).

Cream. Use pasteurised cream, over 40 per cent butterfat. Freeze in cartons (2.5cm/1in headspace). High quality storage life: 6 months.

Cream Cheese. Best blended with heavy cream and frozen as a cocktail dip in waxed tubs or rigid plastic containers. High quality storage life: 3 months.

Eggs. Do not freeze eggs in their shells. Blend lightly with a fork. Add ½ tsp salt or ½ tsp sugar to eggs. Pack in waxed or rigid plastic containers. Label with number of eggs and 'salt' or 'sugar'. Use as fresh eggs as soon as thawed; 3 tbsp whole egg = 1 fresh egg. Egg whites and yolks can be frozen separately. High quality storage life: 12 months.

Milk. Freeze homogenised milk in cartons (2.5cm/1in headspace). High quality storage life: 1 month.

Whipped Cream. Use 1 tbsp sugar to 600ml/1 pint cream. (a) Freeze in cartons (2.5cm/1in headspace). (b) Pipe in rosettes, freeze on open trays and pack in boxes. High quality storage life: 6 months.

• COOKED DISHES •

• Soups

Besides completed soups, meat, chicken and fish stock can all be frozen to use as a basis for fresh soups. These stocks should be strained, cooled and defatted, and packed into cartons with headspace. They are best thawed in a saucepan over low heat.

Soup which is thickened with ordinary flour tends to curdle on reheating, so cornflour is best as a thickening agent. Rice flour can be used, but makes the soup glutinous. Starchy foods such as rice, pasta, barley and potatoes become slushy when frozen in liquid, and should only be added during the final reheating after freezing. It is also better to omit milk or cream from frozen soups, as results with these ingredients are variable.

◆ Cooked Meat and Poultry

Pre-cooked joints, steaks and chops do not freeze successfully, since the outer surface sometimes develops an off-flavour, and reheating dries out the meat. Fried meats also tend to toughness, dryness and rancidity when frozen. Cold meat can be frozen in slices, with or without sauce. Good cooked dishes for freezing include casseroles, stews, cottage pie, galantines and meat loaves, meat balls, meat sauces, and meat pies. It is very important that all cooked meats should be cooled quickly before freezing. Where ingredients such as meat and gravy are to be combined, they should be thoroughly chilled separately before mixing.

It is preferable to freeze meat and poultry slices in gravy or sauce so that they keep their juiciness. Both the meat and the gravy or sauce should be cooled quickly, separately, before packing. These slices are best packaged in foil containers, covered with a lid, and this can save time in reheating as the container can go straight into the oven, keeping the meat moist. These frozen slices in gravy should be heated for 30 minutes at 180°C/350°F/gas mark 4. High quality storage life: 2 months.

For defrosting by microwave, follow manufacturers' instructions.

◆ Casseroles

When making a casserole, it is good sense to double the quantity, using half when fresh and freezing the second half. For freezing, vegetables should be slightly undercooked in the casserole; onions, garlic and herbs should only be used sparingly, or should be added during reheating; sauces should be thickened with tomato purée, vegetable purée or cornflour, to avoid curdling on reheating. High quality storage life: 2 months.

◆ Galantines and Meat Loaves

Galantines are most easily used if cooked before freezing, ready to serve cold. They can be prepared directly in loaf tins, then turned out, wrapped and frozen. Meat loaves can be frozen uncooked. High quality storage life: 1 month.

◆ Pâtés

Pâtés made from liver, game or poultry, freeze extremely well. They can be packed in individual pots ready for serving, or cooked in loaf tins or terrines, then turned out and wrapped in foil for easy storage. Pâtés containing strong seasoning, herbs or garlic should be carefully overwrapped, otherwise they may affect the flavour of neighbouring foods in the freezer. High quality storage life: 1 month.

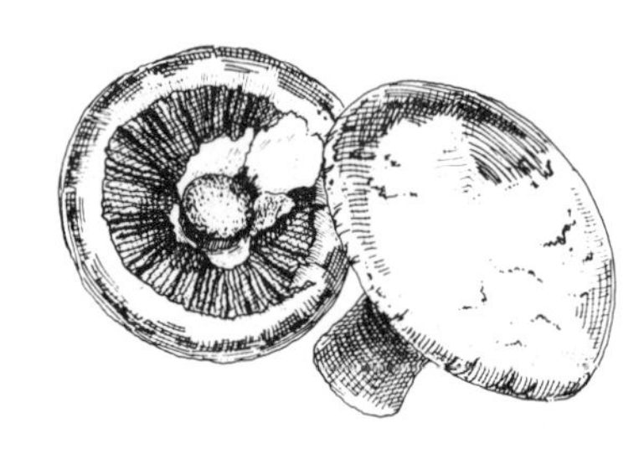

• Cooked fish

Leftover cooked fish can be frozen in the form of a fish pie, fish cakes, or a ready-to-eat dish in sauce. High quality storage life: 2 months.

• Pasta

Pasta can be frozen successfully to be used with a variety of sauces. Composite meals such as macaroni cheese can also be frozen when cooked. Pasta shapes can be frozen to use with soup; but they should not be frozen in liquid as they become slushy.

Pasta should be slightly undercooked, in boiling salted water. After thorough draining, it should be cooled under cold running water in a sieve, then shaken as dry as possible, packed into polythene bags, and frozen. Composite dishes can be reheated in a double boiler or in the oven under a foil lid. High quality storage life: 1 month.

• Sauces

Sweet and savoury sauces can be frozen successfully. They can be in the form of complete sauces such as a meat sauce to use with spaghetti or rice, or you can freeze a basic white or brown sauce to be used with other ingredients when reheated. Sauces for freezing are best thickened by reduction or with cornflour. Mayonnaise and custard sauces cannot be frozen, since the ingredients freeze at different temperatures and give unsatisfactory results.

Sauces can be stored in large quantities in cartons, or in 'brick' form using loaf tins. Small quantities can be frozen in ice-cube trays, then wrapped individually in foil and packed in quantities in bags for easy storage.

• Puddings

A wide variety of puddings can be frozen and this is a way of storing surplus fruit in a convenient form. Puddings which can be frozen include ice cream and pies, pancakes and sponge-cakes which can be combined quickly with fruit, cream or sauces to make complete puddings. Steamed puddings can also be frozen, together with fruit crumbles, gelatine sweets, cold soufflés and mousses and cheesecakes. Milk puddings do not freeze well.

It is better not to put jam or syrup in the bottom of puddings before cooking, as they become soggy on thawing, but dried fruit, fresh fruit and nuts can be added. Highly-spiced puddings may develop off-flavours.

Many cold puddings involve the use of gelatine. When gelatine is frozen in a creamy mixture, it is entirely successful, although clear jellies are not recommended for the freezer. The ice crystals formed in freezing break up the structure of the jelly which becomes granular and uneven and loses clarity. This granular effect is masked in such puddings as mousses.

Suet puddings containing fresh fruit can be frozen raw or cooked. It is more useful, however, to cook them before freezing, since only a short time need then be allowed for reheating before serving. Puddings made from cake mixtures, or any traditional sponge or suet puddings, can also be frozen raw or cooked. Cake mixtures can be used to top such fruits as apples, plums, gooseberries and apricots. These are just as easily frozen raw since the complete cooking time in the oven is only a little longer than reheating time. This also applies to fruit puddings with a crumble topping.

It is useful to use some fruit to make prepared puddings for the freezer. Fruit in syrup can be flavoured with wine or liqueurs and needs no further cooking; this is particularly useful for such fruits as pears and peaches which are difficult to freeze well in their raw state.

Cheesecake. Make baked or gelatine set cheesecake in cake tin with removable base. Cool. Freeze, without wrapping, on a tray. Pack in a box to prevent damage. High quality storage life: 1 month.

Flans (savoury and sweet). Prepare and bake flan, and finish completely. Freeze on a tray, without wrapping. Wrap in foil or polythene, or pack in box to prevent damage. High quality storage life: 2 months (fresh filling); 1 month (leftover meat or vegetables).

Fruit crumble. Prepare fresh fruit with sugar in a foil dish. Cover with crumble topping. Cover with a lid or pack in a polythene bag. High quality storage life: 2 months.

Fruit pies. Avoid using apples, which tend to discolour. Brush bottom crust with egg white, to prevent sogginess. (a) Bake pie. Cool and cover with foil or pack in polythene bag. (b) Cover uncooked fruit and sugar with pastry. Pack in foil or polythene bag. High quality storage life: (a) 4 months; (b) 2 months.

Fruit puddings (suet). Use plums, gooseberries or rhubarb fillings with suet crust. Apples tend to discolour. Prepare in foil or polythene basin and steam. Cool and cover with foil. High quality storage life: 2 months.

Mousses and cold soufflés. Prepare in serving dishes if these are freezer tested. High quality storage life: 1 month.

Sauces (sweet). (a) Sauces made from sieved fresh or stewed fruit freeze well. Pack in boxes in usable portions, leaving headspace. (b) Pudding sauces made from fruit juice, chocolate, etc. can be frozen. They should be thickened with cornflour. Pack in boxes in usable portions, leaving headspace. High quality storage life: (a) 12 months; (b) 1 month.

• Baked goods

Small cakes, buns and rolls are most easily frozen in polythene bags; small iced cakes are better packed in boxes. Large quantities of small, iced cakes can be frozen in single layers, then packed in larger boxes with freezer film or greaseproof paper between the layers. Bread and large cakes can both be frozen in polythene bags. Icings for freezing are best made with butter and icing sugar; cakes should not be filled with boiled icing or with cream as these will crumble on thawing; so will icings made with egg whites. Flavourings must always be pure, as synthetics develop off-flavours in storage (this is particularly important with vanilla and only pure extract or vanilla sugar made with a pod should be used).

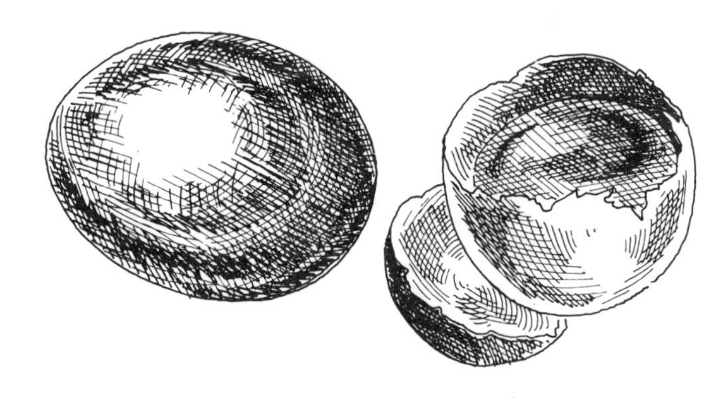

◆ UNCOOKED YEAST MIXTURES ◆

It is possible to freeze unbaked bread and buns for up to two weeks, but proving after freezing takes a long time, and the final texture may be heavier. If unbaked dough is frozen, it should be allowed to prove once, and either shaped for baking or kept in bulk if storage in this form is easier. Brush the surface with a little olive oil or unsalted melted butter to prevent toughening of the crust, and add a little extra sugar to sweet mixtures.

The dough should be thawed in a warm place quickly. Speed will help to give a light textured loaf. After thawing the dough can be shaped and proved again before baking.

◆ Biscuits

Biscuits are the exception to the rule that cooked frozen goods are better than uncooked ones. Baked biscuits do freeze well, but they store equally well in tins, so there is no advantage in taking valuable freezer space for them. The most useful and time-saving way of preparing biscuits is to freeze batches of any favourite recipe in cylinder shapes, wrapped in freezer paper, polythene or foil. The dough will be all the better for having been frozen, giving light crisp biscuits. To use, leave in freezer wrapping in the refrigerator for 45 minutes until just beginning to soften, then cut into slices and bake.

◆ Icings and fillings

Cakes for storage should not be filled with cream, jam or fruit. Butter icings are best, but an iced cake must be absolutely firm before wrapping and freezing. Brief chilling in the refrigerator will achieve this in hot weather. Wrappings must be removed before thawing to allow moisture to escape and to avoid smudging the icing.

◆ Flavourings and decorations

Flavourings must be pure for all icings and fillings, and vanilla extract or vanilla sugar should be used when vanilla is needed. Highly spiced foods may develop off-flavours, so spice cakes should not be used for freezing. Chocolate, coffee and fruit cakes freeze very well.

◆ Pastry

Short pastry and flaky pastry freeze equally well either cooked or uncooked but a standard balanced recipe should be used for best results. Pastry can be stored unbaked or baked; baked pastry keeps longer (baked: 6 months; unbaked: 4 months), but unbaked pastry has a better flavour and scent, and is crisper and flakier.

Pastry can be rolled, formed into a square, wrapped in greaseproof paper, then in foil or polythene for freezing.

Flan cases, pastry cases and vol-au-vent cases are all useful to keep ready-baked. For storage, it is best to keep them in the cases in which they are baked or in foil cases. Baked cases should be thawed in their wrappings at room temperature before filling. They can be heated in a low oven if a hot filling is to be used. For defrosting by microwave, follow manufacturers' instructions.

• Sandwiches

Every filling keeps for a different length of time, so the best general rule is not to store any sandwiches in the freezer for longer than 4 weeks. Sandwiches should be packaged in groups of six or eight rather than individually. An extra slice or crust of bread at each end of the package will help to prevent them drying out.

Avoid fillings which contain egg white, which becomes dry and tough with freezing. Also avoid raw vegetables such as celery, lettuce, tomatoes and carrots, and salad cream or mayonnaise which will curdle and separate when frozen and soak into the bread when thawed. To prevent fillings seeping through, butter the bread liberally. This is easier to do if the bread is one day old.

Sandwiches should be defrosted in their wrappings at room temperature for 4 hours. Follow manufacturers' instructions for defrosting by microwave.

◆ HIGH QUALITY STORAGE LIFE ◆

Item	*High quality storage life number of months*
◆ Meat	
beef	12
ham and bacon (whole)	3
ham and bacon (sliced)	1
lamb	9
minced beef	2
offal	2
pork	6
sausages and sausage meat	1
veal	9
◆ Poultry	
chicken	12
duck	6
giblets	3
goose	6
poultry stuffing	1
turkey	6
◆ Game	
feathered game	10
hare	6
rabbit	6
venison	12
◆ Fish	
oily fish (herring, mackerel, salmon, trout)	2
shellfish	1
white fish (cod, haddock, plaice, sole)	6

Item	*High quality storage life number of months*
◆ Vegetables	
asparagus	9
beans	12
brussels sprouts	10
carrots	10
fresh herbs	10
part-fried chips	4
peas	12
spinach	12
tomatoes	6
◆ Fruit	
apricots	6
cherries	7
currants	10
fruit juices	9
fruit purées	15
gooseberries	10
melon	9
peaches	6
plums	12
raspberries	12
rhubarb	12
strawberries	12
◆ Dairy produce	
double cream	6
eggs	12
fresh butter	6
hard cheese	3
ice cream	3
salted butter	3
soft cheese	6

Item	*High quality storage life number of months*
• Baked goods	
baked bread, rolls and buns	2
breadcrumbs	3
Danish pastry	1
decorated cakes	3
fried bread shapes	1
fruit pies	6
meat pies	3
pancakes (unfilled)	2
pastry cases	3
pizza	1
plain cakes	6
sandwiches	2
savoury flans	2
unbaked biscuits	4
unbaked bread, rolls and buns	2
unbaked cakes	2
unbaked pastry	3

Item	*High quality storage life number of months*
• Cooked dishes	
casseroles and stews	2
curry	2
filled pancakes	1
fish dishes	2
meat in sauce	2
meat loaf	1
pâté	1
roast meat	1
sauces	2
soufflés and mousses	2
soup	2
sponge puddings	3
stock	2

SALTING & SMOKING

Meat and some vegetables may be successfully salted at home, and this is a very early method of food preserving which is enjoying a revival. Care is needed to produce delicious results, but the processing can be worthwhile for those who like to experiment. Some meats can also be successfully smoked at home as well as fish although larger items obviously need larger equipment.

• SALTING VEGETABLES •

• Salt beans

Allow 500g/1lb salt to 1.5kg/3lb beans, but be sure to use block kitchen salt or bag salt recommended as kitchen salt, as table salt contains chemicals which make it unsuitable for preservation. Also be sure to follow quantities exactly for successful bean preservation, and pack into glass or stoneware jars.

Use young, fresh, tender beans, wash and dry them thoroughly and remove strings. French beans may be used whole, but runner beans must be sliced. Put a layer of salt in the jar and then a layer of beans. Continue with alternate layers, pressing down the beans well and finishing with a layer of salt. Cover and leave for a few days until the beans have shrunk and the jar needs topping up. Put in more layers of beans and salt, finishing with a layer of salt. The salt will draw moisture from the beans and become a strong brine. Cover with a moisture-proof lid such as a cork or plastic material tied down tightly. Do not stand stoneware jars on stone, brick or concrete floors as moisture from these materials will be drawn up.

To use the beans, remove some from the jar and wash thoroughly in several waters. Soak for 2 hours in warm water but do not soak overnight, or the beans will toughen. Cook in boiling water without salt until the beans are tender.

If the beans become slimy and do not keep, it is because insufficient salt has been added, or the beans have not been pressed down enough in the storage jar and air pockets have formed, allowing bacteria or moulds to develop.

• SAUERKRAUT •

6 large white cabbages

salt

cooking apples

Trim the outer leaves from the cabbage and cut the cabbage into quarters. Take out the hard stalks and shred the cabbage finely. Pack the cabbage into a large barrel or crock with salt and apples. Do this by putting in a layer of cabbage about 10cm/4in deep and sprinkling with 75g/3oz salt and 1 finely chopped apple. Press down the layers and continue, pressing and pounding very thoroughly so that moisture is released and the cabbage becomes covered with brine. Fill the crock, but leave enough space for the cabbage to swell and ferment without overflowing. Cover with a layer of whole cabbage leaves and a thick layer of salt. Put on a clean cloth and a piece of wood which fits into the top of the crock, and put a heavy weight on top. Keep in a warm place to ferment. After two weeks the level of liquid will have risen considerably. Skim off scum and replace with a fresh cloth. Wash the lid and sides of the crock and replace the lid and weight. Put in a cool place for storage. Put on a clean cloth, and wash the lid and crock every week. The sauerkraut is ready for use in 3–4 weeks when fermentation has ceased.

If sauerkraut is not to be used as soon as it is ready, the juice should be drained off and heated to boiling point. Add the shreds of cabbage and bring to simmering point, then pack at once into heated preserving jars. Adjust the lids and process in boiling water for 25 minutes. Take out the jars, tighten the lids and cool before storage.

◆ SALTING MEAT ◆

Meat may be cured in a wet brine or by a dry cure. It may then be used at once, or frozen for a short period, or smoked.

◆ Ingredients for salting

The chief ingredients for salting are salt, sugar and saltpetre, but fragrant spices and herbs such as juniper berries, peppercorns and allspice can be added to give special flavour to the meat.

Salt must not be table salt which contains chemicals unsuitable for the brining purpose, but should be block salt or bag cooking salt.

Saltpetre (potassium nitrate) is the ingredient which makes salt meat look pink, and should be used as otherwise the meat looks grey and unappetising. Too much saltpetre will harden the meat. It can be bought by the 25g/1oz at chemists.

Sugar for curing may be of the soft brown variety, or demerara, and this gives a good flavour as well as counteracting the hardening effect of the saltpetre.

Herbs and spices give delicious individual flavours to meat cures. The spices should be bought whole and freshly ground when needed, and the herbs are best freshly picked. Juniper berries may be found in many grocers, along with whole allspice, peppercorns and cloves which are the spices most commonly needed for curing.

◆ Storage of cured meats

Meat in a wet brine remains in pickle from 7–21 days, according to type. For longer storage, the meat may be frozen for up to 4 weeks, but the reaction of fats and salt will cause rancidity and off-flavours after that time. Hams may be dried and then smoked, and can then be hung in a cool, dry and airy place, wrapped in muttoncloth; or hams can be wrapped in muttoncloth or linen and stored in a box of wood ash, raised off the floor in a cool dry place.

It is most important to protect cured meat from flies at all stages of curing and storage, or eggs may be laid and maggots can quickly develop.

BRINE FOR TONGUE, BEEF, PORK • OR MUTTON •

750g/1½lb cooking salt

125g/4oz soft brown sugar

15g/½oz saltpetre

2 litres/4 pints water

Dissolve the salt, sugar and saltpetre in the water, bring to the boil and then simmer for 20 minutes. Strain and leave until cold in a large bowl (a large rectangular washing-up bowl is useful for this purpose). Put in the fresh meat and put a plate on top to keep the meat underneath the liquid. Keep in the refrigerator or a very cold larder. Turn the meat every second day. Beef brisket or silverside (1.5kg/3lb) will take 7–10 days; pork or mutton shoulder, leg or loin (1.5kg/3lb) will take 7–10 days; ox tongue (1.5kg/3lb) will take 18–21 days. Small thin pieces of pork such as belly may be brined for 3–6 days.

• BROWN SUGAR • CURE

500g/1lb soft brown sugar

125g/4oz saltpetre

1.5kg/3½lb cooking salt

125g/4oz juniper berries

125g/4oz black peppercorns

This cure will be enough for a 9kg/20lb ham. Remove thigh bone and drain excess blood and liquid. Rub in the brown sugar and leave to stand until the next day. Rub with the saltpetre and scatter over half the salt. Leave until the next day. Grind the juniper berries and pepper and mix with the remaining salt. Put on the ham and rub each day for 1 month. Dry and smoke.

•DRY CURE FOR HAM•

½ lemon

1kg/2lb cooking salt

75g/3oz saltpetre

175g/6oz soft brown sugar

40g/1½oz peppercorns

40g/1½oz mixed spice

25g/1oz cloves

25g/1oz juniper berries

20 bay leaves

20 sprigs thyme

This is a French cure which gives a deliciously aromatic ham. This quantity of ingredients will be enough for 7–10kg/14–20lb ham. The ham should come from a freshly killed pig and the thigh bone must be cut out. Leave the knuckle bone so that the ham can be hung up. Beat the ham so that any blood or juice runs out from where the bone has been removed. Hang the ham in a dark, airy and cool place for 3 days, wiping it well each day with a clean cloth, so that all surplus liquid drains out – this is most important to the success of the cure. Rub the ham with the cut side of the lemon at the end of this time, and prepare the flavoured salt. Mix the salt, saltpetre and sugar together. Put all the spices and herbs into an electric blender and crush them finely. Mix in with the salt. Put a layer of this mixture in a clean dry bowl or crock and put in the ham, skin side down. Pack the rest of the salt mixture round the sides and over the top and be sure that some of the mixture is pushed into the thigh-bone cavity and down the knuckle bone. Put a clean pastry board on top with a weight so that the meat is held underneath the salt. Leave for 10–12 days in a cool dry place (3–7°C/38–45°F). If in doubt about the temperature, keep the ham at the bottom of the refrigerator.

The ham may be cooked at once, after 10 days in the salt mixture. It can also be washed and scrubbed with a nailbrush, then hung to dry for 10 days in a steady drying temperature between 10–15°C/50–60°F but must be protected from flies. The ham may also be dried for 3–4 days and then smoked.

• HAMBURG • PICKLE

4 litres/8 pints water
1kg/2½lb cooking salt
500g/1lb coarse sugar
25g/1oz saltpetre
25g/1oz black peppercorns bruised and tied in a bag

Boil all ingredients for 20 minutes. Clear off scum, leave in a large bowl to cool. When cold, put in meat and keep submerged. A 4kg/8lb joint needs about 14 days.

• TREACLE PICKLE •

4 kg/8lb cooking salt
1kg/2lb black treacle
15g/½oz saltpetre

Remove thigh bone and drain excess blood and liquid. Take half the salt, rub well into the ham and leave for 2 days. Brush on surplus salt, black treacle and saltpetre, and rub well into the ham. Leave for 14 days, rubbing in the mixture daily. Dry and smoke.

• CLEANING AND DRYING •

When meat is taken from the salt cure, it needs preparation for long-term storage if it is not to be cooked at once. The surface should be thoroughly washed in cold water to take off the outer layer of salt, or the meat will become too salty, and there will possibly be a damp surface which can become slimy. If the meat has been dry-cured, it can be simply brushed and left in a cool place so that the salt and saltpetre will continue to do their work. This can be accomplished at the bottom of a refrigerator.

Salt meat does not have to be smoked, but can simply be dried. It is important that the place used for drying is really dry as dampness is attracted by surface salt. The temperature should be about 15°C/60°F. During drying, a white surface layer of crystalline salt will appear which should be carefully removed from time to time.

The dried meat can be stored, and if possible, the storage place should be dark, and again a suitable temperature is about 15°C/60°F. Temperature may be as low as 10°C/42°F, but it is important to avoid variations in temperature as this can cause condensation and the deposit of a film of moisture on the meat. The bacon or ham can be hung up without covering, but it is better to wrap it in paper or cloth and then surround it with a substance which will take up moisture. Charcoal, wood ash or oatmeal (free from mites) are the most easily obtained suitable materials and will not affect the flavour of the meat, but will keep it at an even temperature, dry and free from light.

To store in this way, sift a layer of clean dry wood ash into a box and place the wrapped meat on top. Continue with layers of ash and wrapped meat if there are many pieces to be stored. Finish with a layer of ash about 20cm/8in deep, and keep in a dry place with the box raised on some bricks to prevent dampness rising. If the ashes do get damp, they must be removed, dried and replaced. If oatmeal is used, prepare in the same way, but be sure the storage area is mouse-proof.

If it is preferred to hang up the meat, wrap it in clean linen or calico and wash the bags three times in a solution of lime.

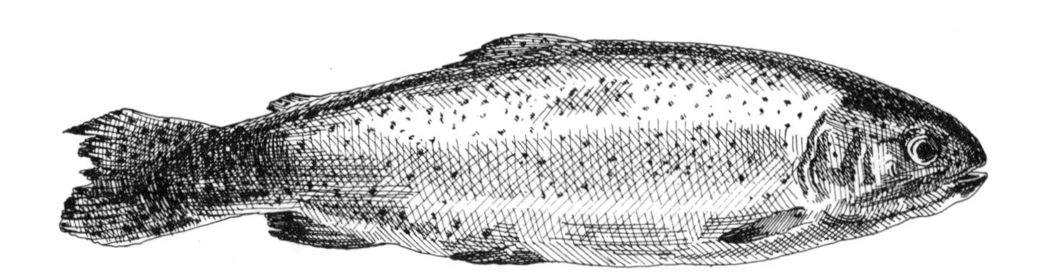

Pear Ginger Jam (page 90).

Granulated Sugar
Pear Ginger Jam

• SMOKING •

The object of smoking is to provide an alternative method of drying, while at the same time introducing preservatives and flavouring. Smoking can retard the onset of rancidity. Smoking must be carried out at a temperature below 32°C/90°F or the fat will melt, and it is important that sufficient moisture is removed from the meat. Smoking should be carried out in a smokehouse or large smoke box, and many butchers and fishmongers will extend this service to good customers who have already prepared their meat. It is possible to construct a smokehouse at home and small home smokers are also available which give a delicious smoked flavour to all sorts of food, but do not smoke food long enough or thoroughly enough for storage purposes.

• Small home smokers

These useful little boxes can be used for smoke-cooking fish, poultry and meat. The meat is placed on a carrier rack over a baffle plate which rests in a metal box containing burning methylated spirits which heats sawdust. The smoke is contained under a lid.

Small home smokers should be placed outside on a level fire-proof base away from inflammable material. They should be kept away from very windy conditions or the effectiveness of the burners will be reduced and it may be necessary to refill them. The amount of smoke created will depend on the quantity of wood dust used, more wood dust producing heavier smoking. The smoke-cooking process will last about 20 minutes, and no pre-cooking is necessary.

Fish may be smoked up to weights of 750g/1½lb, but it is best to start with specimens about 350g/12oz. Two or three fish may be prepared in a small smoker, and up to eight in a larger version. Trout, salmon, mackerel, and herrings are all suitable for smoking. Clean the fish, scale, and remove head and tail if liked, and dry off excess moisture. Do not salt the skin as this will make it peel during cooking. The fish will take about 20 minutes to prepare. Fish roe may be placed on the baffle plate under the fish and will cook in the juices which run from the fish.

Poultry pieces and game can be smoke-cooked. Chicken drumsticks are delicious done this way, and so are pieces of goose and turkey breast. Pieces should weigh about 150g/5oz. The meat may be prepared with seasoning before smoking, or rubbed with lemon and herbs.

Pork chops or lean slices up to 3.75cm/1½in thick and up to 250g/8oz in weight may be smoke-cooked, spread with a little crushed garlic and sprinkled with rosemary. Some sliced onions may be put on the baffle plate under the chops to cook during smoking. *Pork or beef sausages* can also be smoke-cooked without pre-cooking.

Freezer Strawberry Jam (page 94).

• A simple smokehouse

While is is comparatively easy to salt food at home, smoking is a more complicated business and involves the preparation of a smoke-box. Salting preserves food quite well, but smoking has additional preservative qualities as well as giving the food a special flavour. Foods which can be smoked include eel, salmon, trout, cod's roe, ham and bacon, tongue, turkey, goose, duck and chicken, beef, mutton and home-made sausages; some items need brining before smoking.

The simplest form of smokehouse can be constructed in a 50-litre/10-gallon drum with the bottom cut out and a replaceable top which has a few holes punched in. This is mostly useful for smoking trout or haddock, suspended over the concentrated source of heat and smoke. Haddock should be split, cleaned and beheaded, rubbed inside and out with salt, and left overnight, then dried in the open air for three days.

Trout need not be split or beheaded, but the gut should be removed. The fish should then be suspended by the tail on a rod across the top of the drum, and tied with a piece of wire so that they will be at least 30cm/1ft from the fire. The drum should then be up-ended over the fire – best made between bricks on which the drum stands. The heat must be evenly maintained during smoking – which will take from 9 to 12 hours.

In addition to the basic smokehouse, keep a good supply of wood handy. The object of creating the smoke is to make fumes which solidify the albumen in the meat. This halts decomposition; any special flavour is a bonus. The best smoke is produced by slow combustion by hardwood shavings; oak, beech, and hornbeam are excellent, especially with additional flavour provided by juniper and bay. The addition of thyme, sage or heather makes a good variation detectable by some connoisseurs. Resinous woods should be avoided because they can give an unpleasant flavour to some foods.

A slightly more elaborate smokehouse can be constructed for permanent use. This can be a drum or a packing-case which has been made smoke-proof at the joints with insulating tape or thin strips of metal, standing across a trench. The trench should be about 3m/10ft long, 30cm/1ft deep and 30cm/1ft wide, dug in the direction of the prevailing wind, with old paving stones or sheet iron to roof it over. If the lid of the house is hinged, it will be easier to use, and it should have small holes or a tube inserted as a vent.

To use this type of smokehouse, light a fire in the trench at the end furthest from the box, cover with stones or iron sheets, and open the lid of the box. When the fire is red hot, draw it to the end of the trench and put a load of hardwood sawdust between the fire and the box. Leave a little flue space, cover the trench and seal all spaces with earth. The food can be hung across the box and the lid closed and not opened for 48 hours.

The idea is to build up a progressively denser smoke which dries and flavours the food gradually; if the first smoke is too dense, it will form a dry coating on the food which will not then be penetrated by the later smoking. For this type of smoking, the food should be brined first in a strong brine (with enough salt to float a potato). A small fish need only be brined for 20 minutes, but a salmon needs several days, and more elaborate preparation.

The smoking method which is most likely to appeal today is the one using an old-fashioned farmhouse chimney with equipment which can still be used for hams, sausages and other meats. These must be cured first, and are then best left to dry for two or three days before smoking. The food to be smoked can also be hung above the opening of an old bread oven so that the smoke gently enfolds it.

In the latter case, the temperature should never rise above 32°C/90°F, at which point the fat melts and the meat is spoiled. The fire should only be smouldering, not flaming. On the first day, the meat should be smoked for 30 minutes, then rubbed down with pepper, thyme and chopped bay leaves, which will cling to the fat. After cooling and drying for 48 hours, the meat should be smoked again for one hour. After further drying for 48 hours, and one hour's smoking, a light flavour will be obtained. The meat should then be hung in the chimney for two or three weeks before being stored in a cool dry place. The meat will lose about one-quarter of its weight in smoking.

A wide chimney can be used for the complete smoking, simply with the exposed meat suspended high up on a bar or on hooks on wire, but the fire will need careful attention throughout. Smoke the meat for about an hour a day, allowing a total of at least three weeks for complete smoking.

A better method of smoking is by regulating the smoke-flow; construct a smoking-box for the fireplace, to fit on a wall above the fire. To do this, a sheet of metal has to be fitted across the whole chimney with a piece of piping going through it and the wall into the smoking-box. A second piece of piping should then go out of the top of the box and back into the chimney. Meat in the box needs about two hours' smoking a day for eight consecutive days.

◆ Fish smoking

It is best to salt or brine fish before smoking. Put enough salt into water so that it will float a potato. Allow 20 minutes' brining for small fish, but 2 or 3 hours for larger fish. Smoking time will vary according to the fish – about 8 hours will smoke haddock lightly, but 24 hours for kippers. A salmon may be smoked for only 8 hours if it is to be eaten quickly, but is better with salting and longer smoking for longer storage in a cool larder, not a refrigerator.

A salmon needs to be boned before smoking. Scale the fish first, then lift up some small pieces of skin on each side of the back where the flesh

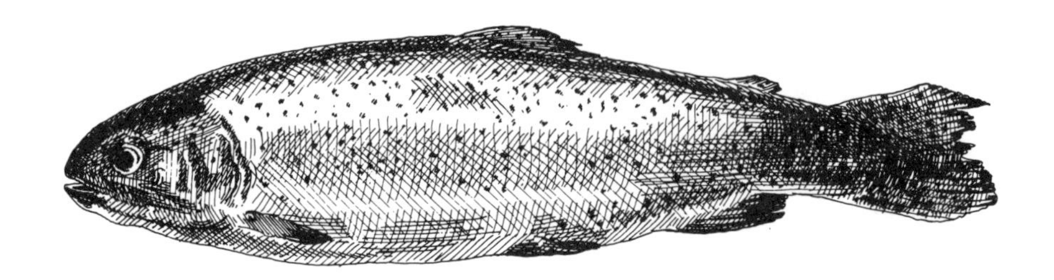

is thickest. Slice these off with a sharp knife, to leave 12mm/½in length diamond-shaped pieces of bare flesh. Rub a pinch of saltpetre into each piece of flesh. Fillet the salmon, cutting on each side of the fins and then remove all bones. Leave the stiff bone under the gill covers to act as a support for hanging. Put salmon sides on a dish and cover with a handful of soft brown sugar and three handfuls of cooking salt. Some people like to rub rum into their salmon too. Turn over the fish next day (the salt will have created a brine) and leave until the following day. Wash in cold fresh water, wipe dry and smoke.

• SAUSAGES •

When meat is being prepared for salting and smoking, or even for freezing, there are often small pieces which are not worth processing whole, but which can usefully be turned into some other kind of delicious food. Sausages are not difficult to make, and sausage-making is a useful way of using up trimmings and small pieces of meat with a variety of flavourings. If a large quantity of sausages is made, they can be frozen and provided a worthwhile addition to family meals.

Sausage-making is not difficult, particularly if a mincer attachment to an electric mixer is used as this will provide large quantities of minced meat quickly, either finely or coarsely ground. A skin-filling attachment is also available for an electric mincer which speeds up the job considerably.

Sausage mixtures are traditionally 'cased', which means stuffing them into a skin or casing, but if skins are not easily obtainable, the sausages may be wrapped in thin sheets of caul fat obtained from the butcher, or the meat may be formed into cylinders or flat patties and used without skins. Caul fat can be obtained in small quantities, usually from specialist pork butchers, but any butcher can order a large supply for you. The fat should be soaked in tepid water, allowing 1 tbsp vinegar to 1 litre/2 pints water. When the fat is soft, it can be cut into any shape and size and the meat rolled in it; the edges should be well overlapped. This gives a firm casing for the meat and an attractive veining of fat.

Sausage casings may be obtained from a friendly butcher in family-sized quantities from his own stock, as commercial supplies are too big for household use. Synthetic casings are easier to handle than natural ones. Natural casings come processed and salted and must be rinsed thoroughly in fresh lukewarm water, then rinsed in cold. When the casings have been rinsed, they must be opened under a jet of water and the easiest way to do this is to turn on the cold water tap, and with the water running, push each length of casing in turn on to the end of the tap. Each length of casing can then be fitted on to the sausage filler ready for the meat mixture. Press out any air which accumulates in the skins as the meat goes into them. If synthetic casings are used, your hands must be dry and free from grease when you are handling them, or the skins will not fill evenly and be under-stuffed. Skins should not be filled too tightly or they will burst when cooked. When the sausage skins are filled, they should be moistened to make it easier to twist them into lengths. If an electric mincer and filler attachment is not available, a small hand filler can be used, similar to a cake icing gun, but it is hard work to force the sausage meat into the casings. It is almost impossible to fill skins without either an electric or hand machine, and the filling will be uneven and rather messy, so it is then better to use caul fat instead.

A particular advantage of making sausages at home is that the meat and flavouring can be varied to suit family taste. The meat may be either coarse or fine and the sausages may be all-meat or with the addition of a small amount of cereal or breadcrumbs. Fresh or dried herbs such as parsley, sage, garlic, rosemary, marjoram and pennyroyal may be used, and such spices as ginger, cloves, coriander or paprika. Start by making a basic sausage-meat recipe and practise filling the skins or wrapping them in caul fat, and then experiment to find the sausages you like best.

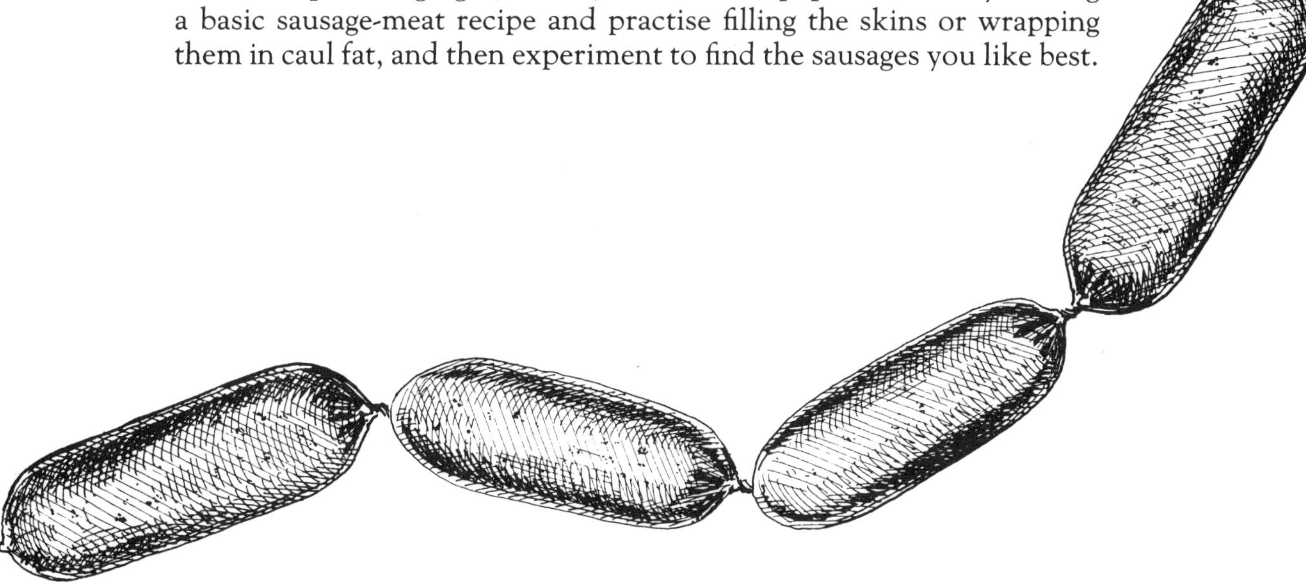

• SAGE SAUSAGES •

- **1.5kg/3lb pork**
- **1 tbsp salt**
- **2 tbsp sage**
- **2 tsp marjoram**
- **¼ tsp ground cloves**
- **¼ tsp black pepper**
- **pinch cayenne pepper**

Use loin or shoulder pork for these special sausages if possible, but make sure there is a good mixture of fat and lean. Cut the meat into cubes and mince finely. Spread the meat out thinly on a board and sprinkle on the salt, finely chopped sage and marjoram, cloves, black pepper and cayenne pepper. Work in the seasoning thoroughly with the hands. Mince the meat twice more. Leave in the refrigerator for 24 hours, then form into sausage shapes or round cakes, or put into skins.

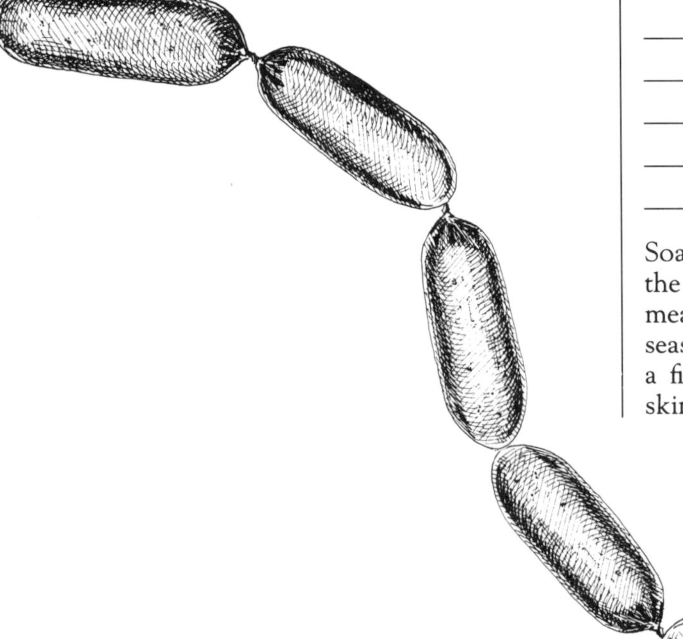

SIMPLE • PORK SAUSAGES •

- **500g/1lb lean pork**
- **250g/8oz pork fat**
- **1 tsp salt**
- **black pepper**
- **½ tsp ground allspice**
- **pinch dried marjoram**
- **25g/1oz dried white breadcrumbs**

Mince the lean pork and pork fat twice. Mix well and season, using salt, freshly ground pepper, allspice and marjoram. Stir in the breadcrumbs. Form into sausage shapes or fill skins.

SPICED • PORK SAUSAGES •

- **750g/1½lb dried white bread**
- **2kg/4lb lean pork**
- **750g/1½lb pork fat**
- **40g/1½oz salt**
- **15g/½oz pepper**
- **8g/¼oz ground ginger**
- **15g/½oz ground mace**

Soak the bread in cold water and wring out with the hands to get rid of excess moisture. Put the meat and fat through a coarse mincer. Mix in the seasonings and bread and put the mixture through a fine mincer. Form into sausage shapes or fill skins.

• BRAWN AND HEAD CHEESE •

Carcasses yield a number of items such as heads and feet which can be prepared for the refrigerator or freezer in the form of brawn or head cheese – pieces of solid meat or shredded meat set in a firm jelly. There are many regional variations in brawns, but basically they are made by simmering the chosen meat very slowly, removing bones and cutting the meat into small chunks or shredding it with forks. The bones are returned to the cooking liquid and simmered to a rich stock which is then mixed with the meat, and pressed down in a bowl. The mixture may be slightly grey in colour but if the meat has been salted, the brawn will be pink. Pale jelly will result from using pork bones for the stock, but if beef bones are added, or some onions in their skins boiled in the stock, the result will be a rich brown colour. Brown colouring may also be introduced with a little vinegar, black treacle or brown sugar. While pork is most commonly used for brawn, beef and lamb are also acceptable, while oxtail, chicken and rabbit are all excellent. Flavours may be varied by the addition of a pinch of nutmeg, a little lemon peel, parsley, thyme, sage or bay.

Meat for brawn may be prepared in a pressure cooker, following manufacturers' instructions.

• BEEF BRAWN •

1 calf's foot

250g/8oz shin beef

8 peppercorns

pinch ground allspice

sprig parsley

sprig thyme

1 bay leaf

1 medium onion

1 clove

pinch of salt

Put the calf's foot into a pan and cover with water. Bring to the boil quickly and strain off the water at once. Wash the calf's foot and put into a clean pan with the beef and cold water to cover. Bring to the boil and skim. Add the peppercorns, allspice, herbs and the onion with the clove stuck in it. Add a pinch of salt and then simmer very gently for 4 hours. Cut up the beef into small pieces with the flesh from the calf's foot. Rinse a basin in cold water and put in the meat. Strain in the cooking liquid and leave to set. Store in the refrigerator up to 3 days, or cover and freeze.

• CHICKEN BRAWN •

1 boiling chicken with giblets

12 peppercorns

1 tsp salt

sprig parsley

strip lemon peel

1 medium onion

½ salted pig's head

2 bay leaves

pinch ground nutmeg

Put the chicken and giblets (except the liver which can be used for another dish) into a pan with water, peppercorns, salt, parsley, lemon peel and unpeeled onion. Simmer for 2 hours. At the same time, put the pig's head into another pan and cover with water. Bring it to the boil and drain. Cover with fresh water and add bay leaves and nutmeg. Simmer for 2 hours. Slice the chicken meat in neat pieces and put some of the pieces into bowls. Put in some chopped meat from the pig's head and then continue with alternate layers of chicken and pork, seasoning each layer with pepper. Mix the chicken and pork cooking liquids, strain and pour over the meat. Leave in a cool place to set. Store in the refrigerator up to 3 days, or cover and freeze.

• JELLIED BRAWN •

½ salted pig's head, with ears and tongue

2 pig's trotters

2 onions

sprig parsley

sprig thyme

1 bay leaf

strip lemon peel

1 blade mace

6 cloves

12 peppercorns

150ml/¼pint wine vinegar

Trim the head and take off the ears and tongue. Trim and clean the trotters. Soak the meat in cold water for 6 hours. Put into fresh water. Do not peel the onions but cut them in half. Add to the pig's head, tongue and ears in the water and bring to the boil. Add a muslin bag filled with herbs, lemon peel and spices. Simmer for 4 hours until the bones slip out easily. Take off the heat and leave to cool slightly. Lift out the meat and bone it. Put the bones back into the liquid and simmer until the stock is reduced to half. Cut the meat including the tongue into small dice. Strain the stock and put the meat back into the stock with the vinegar. Simmer gently until well mixed and slightly reduced. Put the meat into bowls, press down lightly and pour in the stock. Leave in a cool place to set. Store in the refrigerator up to 3 days, or cover and freeze.

NORFOLK • PORK CHEESE •

1 salt pork hock with trotter

pepper

dried sage

Put the meat into a pan and just cover with water. Cover and simmer for 1½ hours until the meat comes off the bones. Take the meat from the bones and cut it into pieces. Toss lightly in freshly ground pepper and powdered sage. Put the bones back into the stock and simmer until the liquid is reduced to half. Strain the stock over the meat, stir well and pour into bowls. When cool and just setting, stir again and then leave to set. Store in the refrigerator up to 3 days, or cover and freeze.

• RABBIT BRAWN •

1 large rabbit

2 pig's trotters

12 peppercorns

blade of mace

2 cloves

Joint the rabbit and soak in cold water. Meanwhile simmer the trotters in water for 2½ hours. Add the rabbit, peppercorns, mace and cloves and simmer for 2 hours until the meat leaves the bones. Cut the meat in pieces and mix the rabbit and the meat from the trotters. Put into bowls and strain in the cooking liquid. Leave in a cold place to set. Store in the refrigerator up to 3 days, or cover and freeze.

• SAGE BRAWN •

½ pig's head

2 pig's trotters

10 sage leaves

salt

pepper

Use an unsalted head for this brawn, and clean the head and trotters carefully. Cover with cold water and add the sage leaves, salt and pepper. Boil gently for 4 hours. When cool enough to handle, take out the bones and chop up the meat. Mix the meat and the cooking liquid and pour into bowls to set. A pinch of ground nutmeg enhances the flavour. Store in the refrigerator up to 3 days, or cover and freeze.

• USING SALTED AND SMOKED PRODUCTS •

Salted and smoked vegetables, meat and fish are all delicious in rather simple dishes because they have a strong character of their own and need little adornment. They are also very useful to liven up otherwise rather dull-tasting but nutritious foods such as pulses (dried beans, peas and lentils) to which they add great succulence.

• BOILED BEEF • & CARROTS

2kg/4lb salt beef (brisket or silverside)
1 bay leaf
sprig parsley
sprig thyme
10 peppercorns
4 onions
12 carrots
For dumplings:
125g/4oz self-raising flour
50g/2oz shredded suet

Soak the meat for 4 hours in cold water and drain well. Put into a large pan with enough cold water to cover. Add the herbs and peppercorns and bring to the boil. Skim, then cover and simmer for $1\frac{1}{2}$ hours. Add the whole onions and the carrots cut in pieces and cook for 15 minutes. Make the dumplings by mixing the flour and suet and mixing to a firm paste with a little water. Roll with the hands into 8 balls and drop into the hot cooking liquid. Cover and continue cooking for 30 minutes. Put the meat on a serving dish, surround with vegetables and dumplings and serve some of the cooking liquid as gravy.

NEW ENGLAND • BOILED DINNER •

2.5kg/5lb salt beef (brisket or silverside)
250g/8oz salt pork
1 cabbage
1 medium turnip
6 carrots
4 parsnips
6 potatoes

Soak the beef and pork for 4 hours. Drain the beef and put into fresh water. Simmer for 1 hour, drain off the cooking liquid and cover the meat with fresh water. Add the pork and continue simmering for 3 hours. Cut the cabbage in quarters, and cut the turnip, carrots and parsnips in neat pieces. Leave the potatoes in their skins. Add to the meat and simmer for 1 hour. Serve the meat surrounded by the vegetables. A boiling chicken may be added to the meat if liked. The meat and vegetables may be served with the cooking liquid, or drained and served with butter and seasoning.

• BOILED TONGUE •

1 salted ox tongue
250g/8oz root vegetables
1 bay leaf
sprig parsley
sprig thyme
6 peppercorns

Soak the tongue for 4 hours and drain well. Put in a large pan and cover with cold water. Bring slowly to the boil and skim well. Add the chopped vegetables, herbs and peppercorns and simmer for about 3 hours until the bones at the root of the tongue will pull out easily. Cool in the liquid, take off the skin and trim away root and bones. Curl the tongue round in a cake tin or straight-sided dish and cover with a plate and a weight. Leave in a cold place for 24 hours before turning out. Cut across in thin slices to serve with salad or in sandwiches.

The tongue may be served hot immediately after cooking. Do not curl it round but skin it and remove root and bones. Arrange flat on a serving dish and cut in thick slices to serve with mushroom sauce, Cumberland sauce, brown gravy or horseradish sauce, and hot vegetables.

BOSTON • BAKED BEANS •

1kg/2lb dried haricot beans
2 tsp mustard powder
1 tsp salt
1 tsp black pepper
50g/2oz tomato purée
1 tbsp golden syrup
1 tbsp black treacle
50g/2oz dark soft brown sugar
2 large onions
250g/8oz salt belly pork
1 tbsp oil

Soak the beans overnight in cold water and drain them well. Mix the mustard, salt and pepper in an oven-proof casserole. Stir in the beans and cover with water. Add the tomato purée, syrup, treacle and brown sugar. Peel and quarter the onions and add to the casserole. Cover tightly and bake at 140°C/275°F/gas mark 1 for 8 hours, adding a little water if necessary so the beans do not dry out. Soak the pork in cold water, and after the beans have been cooking for 6 hours, drain the meat and cut it into 2·5cm/1in cubes. Brown quickly on all sides in the oil, drain and stir into the casserole to cook for the remaining 2 hours.

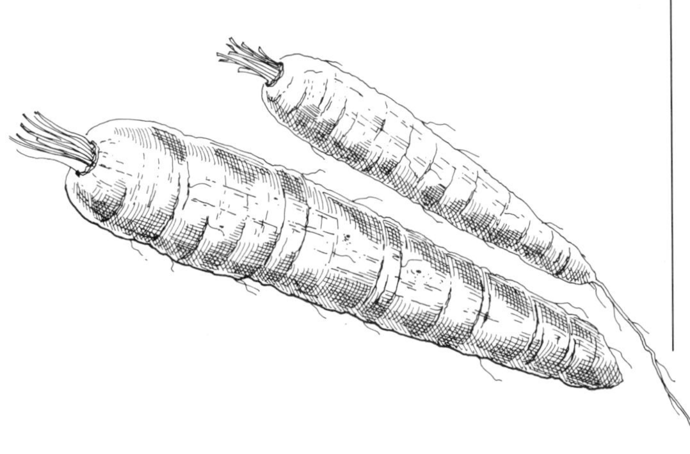

BRAISED • SAUERKRAUT •

1kg/2lb sauerkraut
125g/4oz lean salt pork
2 garlic cloves
black pepper
450ml/¾ pint dry white wine
450ml/¾ pint chicken stock
2 medium onions
4 cloves
Some or all of these may be used:
Frankfurters
smoked boiling sausage
pig's trotters
garlic sausage
baked ham
roast duck or pheasant

Drain the sauerkraut and rinse it well in cold water. Drain thoroughly and squeeze out moisture without breaking the sauerkraut. Put thin slices of salt pork on the bottom of a heavy oven-proof casserole dish. Put half the sauerkraut on top and sprinkle on half the chopped garlic and a liberal seasoning of freshly-ground pepper. Put on remaining sauerkraut, chopped garlic and more pepper. Pour on the wine and chicken stock. Peel the onions and stick the cloves into them. Bury these in the sauerkraut, cover and cook at 170°C/325°F/gas mark 3 for 3 hours so that the liquid is absorbed but the sauerkraut remains firm-textured. Serve with boiled potatoes and a selection of meats, plus a choice of mustards and some rye bread or pumpernickel. The various sausages should be fried or boiled according to type (follow instructions on can, jar or packet); the pig's trotters should be boiled with an onion until tender; the meats and poultry freshly baked, fried or roasted. Each kind of meat should be prepared separately so that the flavours do not become mixed before service.

COLD • PRESSED BEEF •

3·5kg/7lb salt beef (brisket)
2 carrots
1 onion
1 turnip
20 peppercorns
6 cloves

Soak the beef for 4 hours in cold water. Peel and cut up the vegetables roughly. Put into large pan with the drained beef, peppercorns and cloves. Cover and simmer for 6 hours. Lift out the meat, drain it well and wrap in a clean piece of cloth or muslin. Put on to a board and put a flat plate or tin on top. Press down with heavy weights and leave in a cold place for 12 hours. Cut in very thin slices to serve with salad and pickles.

• SALT PORK • WITH CABBAGE

1kg/2lb salt pork or bacon
1 large white cabbage

Keep the meat in one piece and put it into a heavy saucepan. Just cover with water and bring to the boil. Simmer for 30 minutes. Drain off the cooking liquid and reserve, leaving about 150ml/¼ pint in the pan. Wash the cabbage well and cut it into eight sections. Put the cabbage pieces round and on top of the meat. Cover and simmer for 3 hours until the meat and cabbage are tender. Stir the cabbage from time to time so that it does not burn, adding a little of the original cooking liquid if necessary. Serve the pork or bacon in slices with the cabbage.

• SALT PORK • WITH FRIED POTATOES

250g/8oz salt pork
2 tbsp vegetable oil
1kg/2lb boiled potatoes
1 tbsp chopped fresh sage

Use any odd ends of salt pork and slice or chop them roughly. Wipe them well and fry in the oil until the fat has all run out. Lift out the crisp pieces of pork and keep on one side. Slice the potatoes thickly and fry in the fat, adding the sage when the potatoes are golden on one side. Fry the other sides until golden. Drain off the fat and serve the potatoes very hot and sprinkled with the pieces of pork.

◆ To cook a bacon joint

Weigh the joint of bacon. A piece about 2–2.5kg/4–5lb is a good size for a family meal and some cold meals afterwards. Soak in cold water overnight, drain and put in a saucepan covered with cold water. Add a bay leaf and a tablespoon brown sugar with 4 peppercorns. Bring to the boil, cover and simmer, allowing 25 minutes per 500g/1lb and 25 minutes over. Cool in the cooking liquid to eat cold, or eat hot with parsley sauce or Cumberland sauce.

To bake a bacon joint, cook it in water for half the cooking time, drain and strip off the rind. Wrap in foil and bake at 190°C/375°F/gas mark 5 for half the remaining cooking time. Unwrap the foil, score the fat into squares or diamonds and glaze before finishing cooking at 220°C/425°F/gas mark 7. To glaze the joint, either (a) sprinkle with brown sugar and stud the diamonds with whole cloves; (b) sprinkle with sugar and baste frequently with 150ml/$\frac{1}{4}$ pint sweet cider; (c) brush with thick marmalade and garnish with slices of apple dipped in lemon juice. Baste often during cooking, adding a little more marmalade if necessary.

◆ Smoked fish pâtés

Any smoked fish, such as kippers, bloaters, mackerel, trout and salmon, makes delicious pâtés for serving with toast, in sandwiches, or with salad. Kippers and bloaters need to be grilled or poached first, then cooled and flaked. Mackerel, trout and salmon do not need further cooking. Flake the fish and mash it well with its own weight in butter, a little lemon juice and pepper. Press into small pots and chill before serving. For variety, try mashing the fish with a little thick cream, or with full-fat soft cheese. A pinch of cayenne pepper or Tabasco sauce brings out the flavour of the fish.

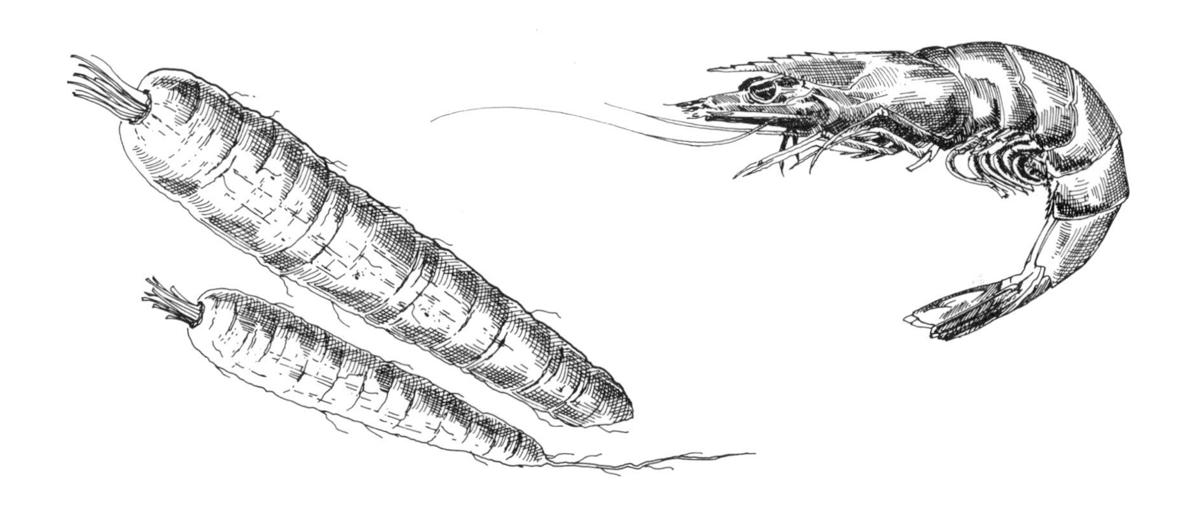

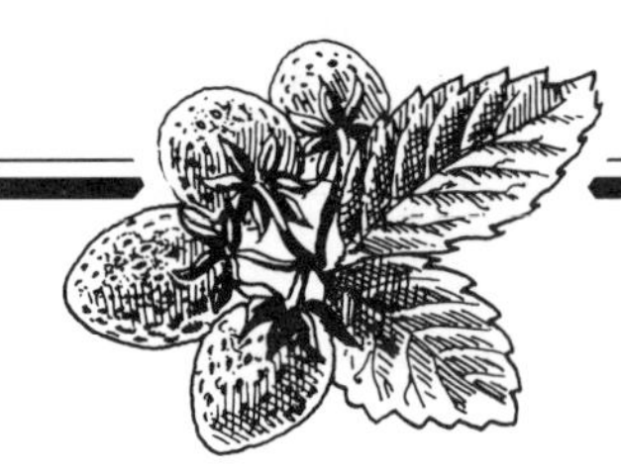

POTTING

Potted meat and fish have been prepared for hundreds of years, preserving cooked flesh under a thick layer of butter or lard which kept out the air and gave a limited storage life in the days before refrigeration. Cheese was often prepared in this way too, and the method was invaluable for making the best of left-over food as well as for preserving larger quantities.

Potted fish and game were popular breakfast dishes, or taken for outdoor meals, and potted meat was a favourite part of a substantial tea, particularly in the Midlands and the North.

The food to be 'potted' was either cooked separately and then mixed with butter and seasonings before being sealed under a lid of fat, or it could be cooked slowly in fat which produced a richly flavoured mixture to be mashed and potted under the preservative layer of extra fat.

The best pots to use are small ones, as once the hard fat lid has been broken, the contents must be eaten quickly. Small china soufflé dishes or ramekins, pâté dishes or straight-sided glass jars are all suitable. The contents must be pressed down tightly to eliminate all air pockets and dishes should be filled very full, just leaving enough space for the layer of sealing fat to be poured on. This layer of fat must be thick enough to seal the food completely.

The meat or fish to be potted should be finely shredded or minced, and then mashed. An electric blender may be used, but only a little can be processed at a time. Without a blender, it is necessary to use a pestle and mortar, or a wooden spoon in a large bowl to mash the ingredients together. Seasoning is important for these foods. The main ingredient may already be salted, but freshly ground pepper is an essential addition. A little ground nutmeg and/or mace bring out the flavour of both meat and fish, and a little ground ginger may also be used. Anchovy essence is useful for potted fish, and a few drops improve the flavour of potted beef.

When the pots have been prepared, they should be stored in the refrigerator and should last up to 3 weeks; they may also be frozen and will last 12 months. Potted meat or fish may be served with hot toast, or fresh wholemeal bread or used as a sandwich filling. Potted fish in particular is good sliced to serve with salad. It is best to bring the potted product to room temperature before serving as the flavour will not be so good when chilled.

The most essential ingredient in potting is clarified butter, which is free from any foreign matter which might allow the fat to become rancid. The clarification process also drives out air and causes the butter to become very solid when it is cold. It is worth preparing a quantity of this butter because it is excellent for frying as it does not blacken and burn and has a good flavour.

◆ To clarify butter

Put *unsalted* butter in a thick saucepan and melt over very low heat, skimming off any foam from the surface. The butter must not brown, but will become very clear with sediment sinking to the bottom of the pan. When foam stops rising to the surface, take off the heat and strain the fat through a double thickness of butter muslin or cheesecloth in a sieve. Leave any excess sediment or milky liquid at the bottom of the pan. Leave the butter to cool to lukewarm before using, and store any surplus in a covered jar in the refrigerator.

BLOATER OR ◆ KIPPER PASTE ◆

2 plump bloaters or kippers

butter

salt

pepper

50g/2oz clarified butter

Bloaters are smoked herring which are a speciality of the east coast, but they are often difficult to find. Grill the fish until just cooked but not dry. Skin them and remove bones. Weigh the flesh and use the same weight of butter. Mash the flesh and butter together and press into pots, seasoning to taste. Cover with clarified butter.

◆ CHICKEN & HAM ◆ PASTE

250g/8oz lean cooked ham

125g/4oz ham fat

50g/2oz butter

pinch ground mace

salt

pepper

250g/8oz cooked chicken

125g/4oz clarified butter

Chop the ham and fat finely and mash with the butter, mace, salt and pepper, to a paste. Chop the chicken and mash it well, seasoning lightly. Put half the ham mixture into a pot and top with half the chicken. Add the remaining ham and top with the remaining chicken. Press down very firmly and cover with clarified butter.

GATESHEAD • POTTED SALMON •

1kg/2lb salmon

salt

pepper

pinch ground nutmeg

pinch ground mace

250g/8oz butter

250g/8oz clarified butter

Scale the salmon and wash it well. Dry and split the fish to remove the backbone. Season with salt, pepper and spices and put into an oven-proof dish with the butter. Leave to stand for 3 hours, then cover and bake at 180°C/350°F/gas mark 4 for 1 hour. Drain and leave to stand until the liquid has run out. Cut in pieces the size of pots and arrange in layers until the pots are filled, putting the skin upwards each time. Put on weights to press the salmon until cold. Cover with melted clarified butter.

• HAM & TONGUE • PASTE

125g/4oz cooked ham

125g/4oz cooked tongue

250g/8oz clarified butter

pinch ground nutmeg

pinch ground mace

pinch ground ginger

pinch thyme

black pepper

Cut the ham and tongue in small pieces and mince finely twice. Mix with 175g/6oz clarified butter, spices, crushed thyme and pepper. Taste for seasoning and add a little salt if necessary. Pack into small pots and cover with remaining butter.

• PORK RILLETTES •

1kg/2lb belly pork

125g/4oz flare fat or lard

salt

pepper

pinch ground nutmeg

Weigh the meat when the skin and bones have been removed. If flare fat is not available from the butcher, use pure lard. Cut the pork and fat into 2.5cm/1in cubes and put into a heavy pan with 150ml/¼ pint cold water. Cover and cook at 140°C/275°F/gas mark 1 for 4 hours. Turn into a sieve so that the liquid runs through, and keep this liquid. Pull apart the meat and fat with two forks so that it is shredded finely, season well and pack into a pot. Strain over the reserved liquid fat. Cool and cover and store in a cold place. Rillettes will keep for months in a cold place. The mixture should be soft enough to spoon out and spread on wholemeal bread or toast. For a richer flavour, add a crushed garlic clove, a bay leaf and a sprig of thyme to the meat while it is cooking, but remove the herbs before shredding the meat.

Rillettes can also be prepared from mixtures of poultry or rabbit with pork. Try using a young rabbit with 350g/12oz belly pork and 125g/4oz flare fat or lard; or a mixture of equal weights of chicken, pork and fat; or use equal quantities of pork, boned goose and lard. Usually, rillettes made with poultry and rabbit are rather smooth, but pork rillettes should be more 'thready'. The meat must always be cooked very slowly so that the pieces of meat do not become hard.

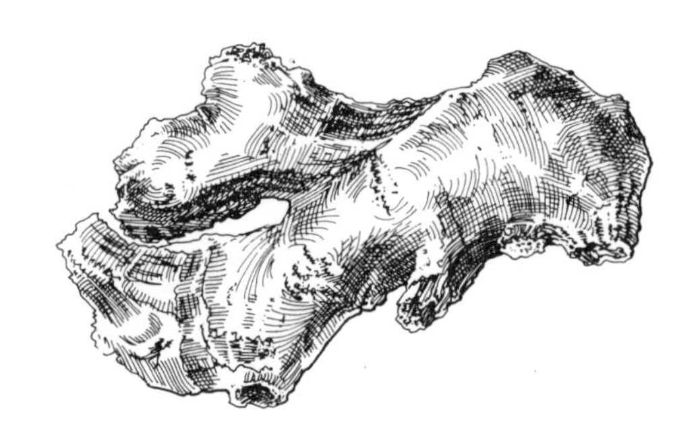

Orange Shred Marmalade (page 97).

◆ POTTED BEEF ◆

500g/1lb shin beef

1 tsp salt

pinch pepper

125g/4oz clarified butter

Cut the meat into very small pieces. Season and put into an oven-proof bowl. Cover with a piece of foil or greased greaseproof paper and put the bowl into a saucepan of water. Cover the pan and simmer over low heat for 2½ hours. Strain the liquid into a bowl and mix with 75g/3oz clarified butter. Mince the meat and then pound until thoroughly soft and smooth. Mix with the liquid and butter. Put into a clean dish and pour on the remaining butter.

POTTED ◆ CHESHIRE CHEESE ◆

500g/1lb Cheshire cheese

75g/3oz butter

1 tsp mustard powder

50g/2oz clarified butter

Grate the cheese and mash it with the butter and mustard. Put into small pots and cover with clarified butter. Another version of potted cheese may be made using the same quantities of cheese and butter, but seasoning with a pinch of ground mace and 4 tbsp dry sherry.

◆ POTTED GROUSE ◆

2 old grouse

1 medium carrot

1 medium onion

50g/2oz streaky bacon

bunch mixed herbs

salt

pepper

125g/4oz clarified butter

Clean the grouse. Slice the carrot and onion finely and chop the bacon. Put the bacon into a pan and heat until the fat runs. Add the carrot and onion and cook gently until golden. Transfer to a casserole and add the herbs, salt and pepper and the grouse. Just cover with stock or water and put on a lid. Cook at 150°C/300°F/gas mark 2 for 2½ hours. Take out the pieces of carrot and onion. Remove the meat from the grouse and put through the mincer with the bacon pieces. Pound to a paste with a little cooking liquid and press into a dish. Cover with clarified butter.

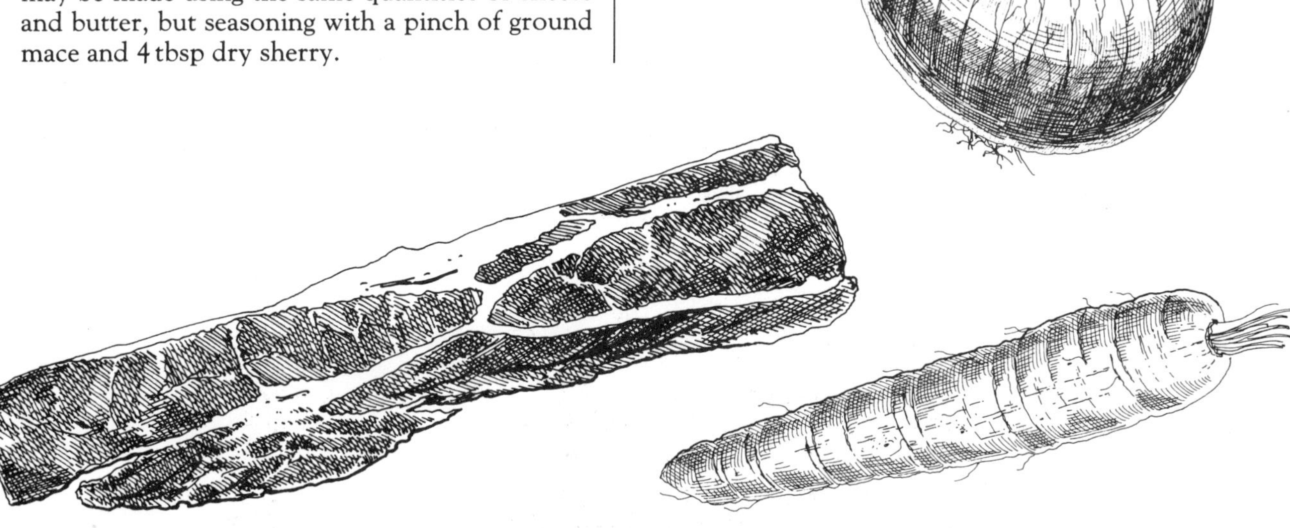

Crab Apple Jelly (page 101).

POTTED HERRINGS

12 herrings

125g/4oz salt

15g/½oz ground nutmeg

15g/½oz ground cloves

8g/¼oz ground ginger

3 bay leaves

4 strips lemon peel

250g/8oz butter

250g/8oz clarified butter

Wash and gut the herrings, and take off the heads, tails and fins. Remove the backbones and roes. Mix the salt and spices and season the fish with this mixture. Put into an earthenware dish and arrange the bay leaves and lemon peel on top. Dot with butter and cover with a piece of foil. Bake at 180°C/350°F/gas mark 4 for 40 minutes. Cool slightly and remove skins. Flake the flesh and press into pots with any cooking liquid. Pour on melted clarified butter. Mackerel, trout and perch may be potted in the same way.

POTTED LOBSTER

500g/1lb lobster meat

250g/8oz clarified butter

pinch ground mace

pinch ground nutmeg

pinch ground cloves

salt

pepper

3 bay leaves

Use freshly cooked lobster and take out the flesh as whole as possible. Put 50g/2oz clarified butter in an oven-proof dish and put the lobster pieces on top. Add the seasonings and bay leaves and top with 50g/2oz butter. Cover and bake at 170°C/325°F/gas mark 3 for 20 minutes. Drain and pack the lobster pieces into a pot. Cool and cover with remaining melted butter. This lobster may be eaten cold or served hot in a sauce.

POTTED MACKEREL

2 mackerel

150ml/¼ pint dry cider

2 bay leaves

1 shallot

2 anchovy fillets

175g/6oz clarified butter

pinch ground nutmeg

pinch ground ginger

pinch cayenne pepper

pinch ground mace

black pepper

Clean the mackerel and put them into an oven-proof dish. Cover with the cider, bay leaves and finely chopped shallot. Cover with a lid or piece of foil and bake at 180°C/350°F/gas mark 4 for 30 minutes. Cool the fish in the liquid. Drain off the liquid and remove the shallot and bay leaf. Skin the fish and flake the flesh. Mix with finely chopped anchovies and 100g/4oz clarified butter and mash to a smooth paste. Add the seasonings and pack into a jar. Pour on the remaining clarified butter.

• POTTED MEAT • OR POULTRY

cooked meat
salt
pepper
pinch ground nutmeg
softened butter
clarified butter

Put any kind of cooked meat or poultry through the fine blade of the mincer twice. Season to taste and mix well with some softened butter. Press into pots and pour over clarified butter. Ham and chicken are particularly good prepared in this way, and useful for sandwich spreads.

POTTED • MUSHROOMS •

500g/1lb field mushrooms
40g/1½oz butter
pinch salt
pinch cayenne pepper
pinch ground mace
50g/2oz clarified butter

Cut the stems from the mushrooms and wipe the caps clean. Put the butter into a thick saucepan and when it has melted over low heat, add the mushrooms. Shake them as they cook gently for 3 minutes. Add the salt, cayenne pepper and mace and continue simmering and shaking until they are very soft. Drain in a colander and leave until cold. Press into pots and pour on melted clarified butter. These mushrooms may be served cold or can be added to other recipes together with the butter in which they are preserved.

POTTED • PARTRIDGE •

2 partridges
salt
pepper
250g/8oz butter
300ml/½ pint stock

Season the birds inside and out with salt and pepper. Put a lump of the butter in each bird, and put the birds in a casserole. Add the stock and remaining butter. Cover with two sheets of foil and the casserole lid. Cook at 170°C/325°F/gas mark 3 for 1½ hours. Leave until cold and the butter will rise and form a seal on top. Do not uncover until they are to be eaten, and keep in a cold place.

POTTED • PIGEON •

3 pigeons
1 tsp Worcestershire sauce
salt
pepper
125g/4oz clarified butter

Skin the pigeons and clean them. Put into a saucepan with just enough water to cover and simmer until the meat leaves the bones easily. Take the meat from the bones and mince it finely. Put the bones back into the cooking liquid and simmer until the liquid is reduced to 300ml/½ pint. Season the minced meat with sauce, salt and pepper. Mositen with the bone stock and a little of the butter. Press the mixture into pots and cover with remaining butter.

POTTED • RABBIT •

1 rabbit
50g/2oz butter
1 sugar cube
1 medium onion
12 cloves
12 allspice berries
6 peppercorns
pinch ground nutmeg
250g/8oz clarified butter
2 tsp Worcestershire sauce

Cut the rabbit into joints and soak them in salted water for 2 hours. Dry well and put the rabbit joints into a casserole. Add the butter, sugar, onion stuck with cloves, allspice, peppercorns and nutmeg. Put on the lid and cook at 150°C/300°F/gas mark 2 for 3 hours. Cool and take the meat from the bones. Put the meat through the mincer twice and mix with the juices from the casserole, half the clarified butter and the Worcestershire sauce. Press into pots and cover with the remaining clarified butter.

• POTTED SHRIMPS •

300g/10oz butter
500g/1lb peeled shrimps
½ tsp ground mace
½ tsp ground nutmeg
pinch cayenne pepper

Take 125g/4oz butter and melt it slowly. Skim off any foam and take the butter from the heat. Spoon the clear butter into a bowl and discard the milky fluid at the bottom of the pan. Melt the remaining butter and stir in the shrimps and spices so that the shrimps are completely covered in butter. Pour into individual dishes or one large one. Leave for about 10 minutes until cool and almost set. Pour on the clarified butter and leave until cold. Keep in a cold place.

POTTED • SMOKED FISH •

6 smoked trout or mackerel
175g/6oz butter
½ tsp lemon juice
salt
pepper
125g/4oz clarified butter

Use plump, freshly smoked fish for this. Remove the skin and bones and flake the fish. Soften, but do not melt the butter. Put the flaked fish, butter, lemon juice and seasoning into an electric blender, and blend until well mixed but not completely smooth. If a blender is not available, mash the fish and other ingredients together with a wooden spoon. Press into pots and cover with melted clarified butter.

POTTED • STILTON CHEESE •

1kg/2lb Stilton cheese
250g/8oz butter
pinch salt
pinch ground mace
2 tbsp port
125g/4oz clarified butter

Cut the cheese into small pieces and put into a bowl with slightly softened butter. Mash together until well blended. Season with salt and mace and moisten with port. Press into small pots and cover with clarified butter.

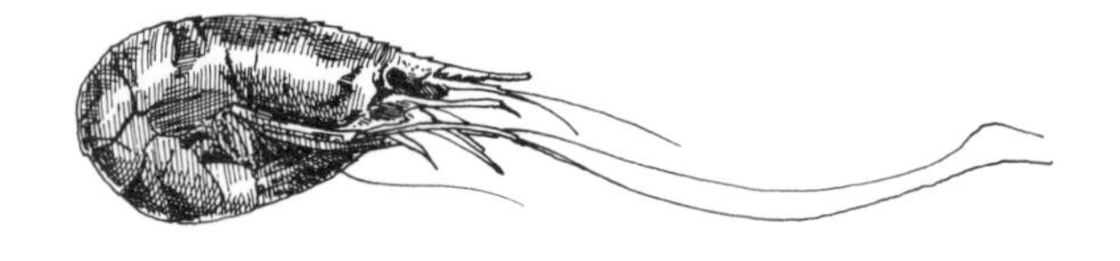

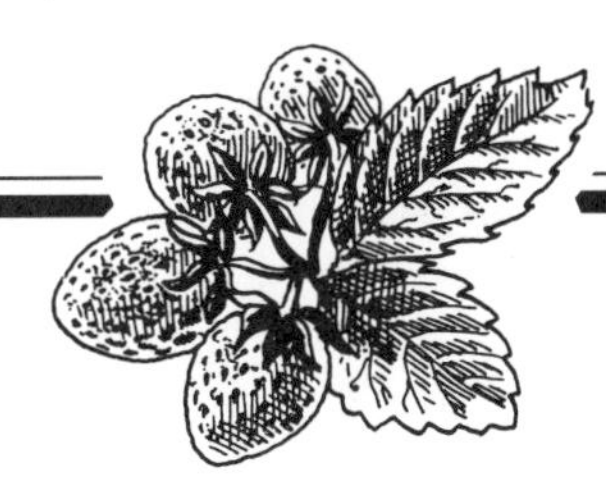

DRYING

Drying is a very old method of preserving, with fish, meat, fruit and vegetables traditionally dried by the action of sun and wind. Today, it is practicable to dry fruit and vegetables at home, if the correct temperature is adhered to and there is adequate ventilation. No expensive equipment is involved, but drying racks can be made.

To make drying trays, nail together four wooden laths in a square and stretch wire gauze or cheesecloth between the framework. Wire trays will need the added protection of loose pieces of cheesecloth or muslin to prevent fruit taking the imprint of the wire mesh. Cheesecloth should be well washed to remove any dressing which will scorch easily. Impromptu racks can be oven racks or wire cake trays covered with muslin.

To dry fruit, use very fresh, ripe specimens as ripe fruit dries more quickly, retains a better colour and has a much better flavour. Prepare the fruit according to its kind, put on the trays and dry at a temperature between 60–65°C/120–150°F. The fruit should be heated slowly at first to prevent the outside hardening, which will prolong the drying process as it will prevent moisture evaporating from the centre of the fruit. If this happens to plums, the skins will burst. The fruit may be dried in the oven, a warm cupboard or on a rack over a stove if there is no moisture. The heat should be constant and gentle with a current of air to carry away moisture, and this is best obtained by leaving the cupboard or oven door slightly open. The fruit should be dried, but not cooked or scorched during processing. After heat drying, remove the fruit from the oven and expose to ordinary room temperature for 12 hours to cool. Pack in wooden or cardboard boxes lined with greaseproof paper, and store in a very dry place, but *not* in airtight containers.

Apple rings and pears. This is a useful way to keep apples if the freezer is full, or there is not the time or equipment for bottling. Pears can be dried, but the juicy varieties are rather difficult to handle.

Apples should be ripe but not over-ripe. Using a silver or stainless steel knife, peel and core the apples and cut into rings about 6mm/¼in thick. Put into a basin of salt water immediately (15g/½oz salt to 1 litre/2 pints water) and leave for 10 minutes. The rings should be threaded on a stick which can rest on the runners of an oven, and the rings should not touch each other. Dry at 65°C/150°F/gas mark ½ for some hours. Apples should be like dry chamois leather, moist and pliable. Cool in the air before packing tightly in paper bags, or in dry jars or tins. Store in a dry, dark place.

It is best to soak fruit for 24 hours before cooking, and to use the soaking water for cooking, flavoured with a little lemon or vanilla, a clove or a piece of ginger.

Cut pears in half and process as for apples.

Dried Grapes. Use fresh, fully ripe fruit, and dry whole. Spread out in a single layer, and test by squeezing. When they are dried, no moisture will appear; 500g/1lb fresh fruit will give about 125g/4oz dried.

Dried Plums. Plums must be ripe. Halve and stone them. Stretch muslin on sticks to replace open shelves, and dry plums on these. Dry slowly, as for apples. Plums can also be dried whole.

Dried Beans. Leave haricot beans on the plants until dry and withered. Pull up plants and hang in an airy shed. Shell the beans and store as for seed in glass jar. Young French and runner beans may be sliced, blanched for 5 minutes, drained and dried in a thin layer.

Dried Mushrooms. These dry very well. Cut the stalks off short and peel. Lay them on the oven racks and when they are dried quite stiff, store them away in glass jars. When preparing them for use soak them in water and simmer in a little stock. Only dry very fresh mushrooms.

Dried Onions and Leeks. These well repay drying, as so often they do not keep. Peel them and remove any bad parts, and slice them on to muslin on oven racks. Leeks may be cut in strips lengthwise if preferred. Move them about occasionally to help the drying, and when they are quite crisp take them out, and after a little while store them away.

Dried Peas. Leave marrowfat varieties on the plants until dry and withered. Pull up plants and hang in an airy shed. Shell the peas and store as for seeds in glass jars. For young, sugary peas, blanch them for 5 minutes, drain them and shake off excess moisture. Spread out in a thin layer and dry very carefully, without scorching.

• FRUIT •

Delicious fruit pastes are simple to make and may be dried to serve as unusual sweetmeats. Dried apples coated in sugar also make a delightful treat.

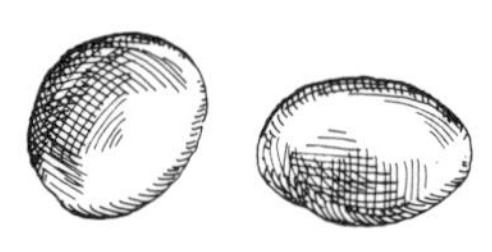

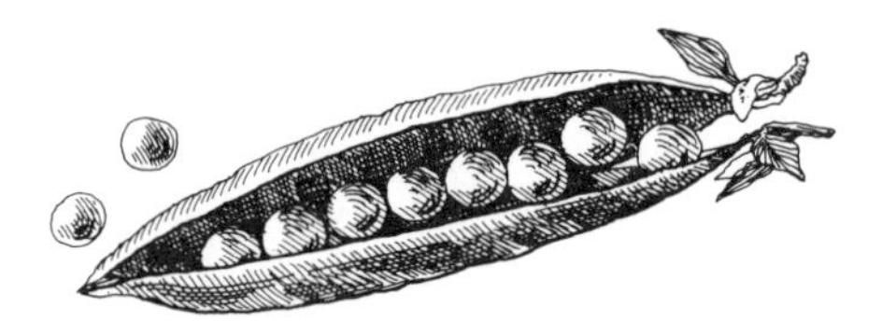

• APRICOT PASTE •

ripe apricots

sugar

caster sugar

Take stones from the apricots and cook the apricots in as little water as possible to prevent sticking. Put through a fine sieve and weigh the pulp. Mix with an equal quantity of sugar and heat and stir until all the moisture has evaporated and the mixture is dry. Roll the paste on a sheet of paper sprinkled with caster sugar, and leave to dry in the sun or in an open oven. The paste should be leathery so that it can be rolled up.

• QUINCE PASTE •

quinces

sugar

icing sugar

Do not peel or core the quinces, cut them into small pieces. Simmer in just enough water to cover until very soft. Sieve the pulp, weigh and take an equal weight of pulp and sugar. Put them into a thick pan and stir over a low heat until the mixture dries and leaves the sides of the pan clear. Cool slightly and then roll out 12mm/½in thick on a board dusted with icing sugar. Stamp out rounds and leave them to dry, turning often until they are the texture of leather. The mixture can be dried out on the rack of a cooker, or in a cool airing cupboard. Dust with icing sugar and store in tins. *Apple paste* and *pear paste* may be prepared in the same way.

REDCURRANT OR • BLACKBERRY PASTE •

1kg/2lb redcurrants or blackberries

sugar

300ml/½ pint water

caster sugar

Heat the currants or berries in water to cover until they burst and are soft. Drain through a jelly bag and weigh the juice. Take an equal quantity of sugar and add to the fruit. Heat slowly, stirring all the time until the mixture is thick and dry. Put the paste into a baking tin and sprinkle with caster sugar. When cold and hard, cut into pieces with a knife, dip in caster sugar and store in a wooden box lined with greaseproof paper.

NORFOLK • BIFFINS •

Norfolk Biffins are red-cheeked apples which are dried out very slowly and are traditional to Norfolk, although a similar product occurs in Normandy. Choose apples which are unblemished and put them on clean straw on a wire cake rack. Cover well with more straw, and put into a very low oven for 5 hours (the lowest oven of an Aga is ideal for this). Take out the apples and press them very gently to flatten them slightly without breaking the skins. Return to the oven for 1 hour, take them out and press them again. When they are cold, coat them lightly with sugar which has been melted over a low heat without colouring. Store in airtight boxes in layers between waxed paper. Any crisp eating apple, such as Cox's Orange Pippins, may be prepared in this way.

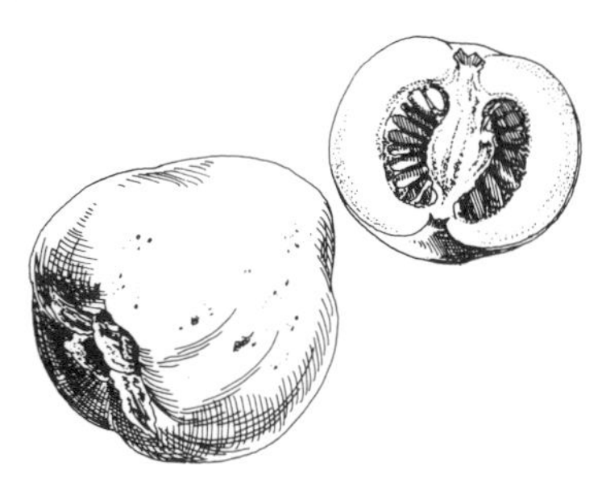

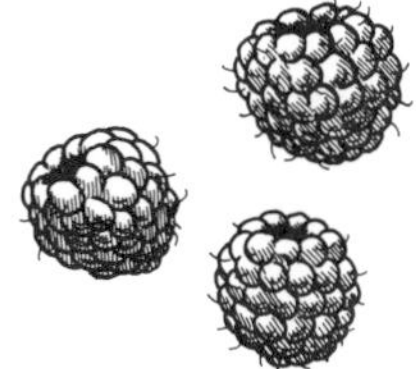

• DRIED HERBS •

Herbs for drying should be picked just before they come into flower. They are best picked on a dry day before the sun is hot. Remove leaves from the stalks of large varieties, and divide the small herbs into sprays. Put on shallow trays in thin layers and leave in a cool oven (50°C/125°F/gas mark ½) or an airing cupboard, or on a cooker rack, until dry. This gives clean, freshly-coloured herbs with good flavour. Parsley is not very easy to dry; it should be placed in a hot oven (200°C/400°F/gas mark 6) for 1 minute, and then be finished off in a cool oven or airing cupboard; then it will retain its bright green colour. When the herbs are dry, remove the stems and rub the herbs down with a rolling pin. Put into small tins or jars, and store in a cool, dry place. Dried herbs should be used within a year, sooner if possible.

Drying Herbs by Microwave. Wash herbs well and pat dry. Put a piece of kitchen paper in the microwave oven and spread out a handful of herbs. Microwave 2–3 minutes on full power, shaking the paper every 30 seconds so that the herbs dry evenly. Leave until cold and then crumble into jars. Drying by microwave retains full flavour and fresh colour.

HERB • SEASONING •

dried orange peel

dried thyme

dried marjoram

dried hyssop

Pound the orange peel to powder. Mix two parts orange peel with one part of each of the herbs. Keep in a well-stoppered jar to use for stuffings, forcemeats and meat loaves.

SWEET HERB • SEASONING •

50g/2oz parsley

50g/2oz marjoram

50g/2oz chervil

25g/1oz thyme

25g/1oz lemon thyme

25g/1oz basil

25g/1oz savory

15g/½oz tarragon

Dry the herbs when they are in season and weigh them when they are dried. Rub each one to a fine powder, and sift through a strainer. Mix together and keep in a tightly-stoppered bottle for seasoning.

◆ SEASONING ◆ SALT

1 tsp mixed dried herbs
15g/½oz mixed ground cloves, mace and ginger
25g/1oz pepper
25g/1oz salt

Mix all the ingredients well together and rub through a sieve. Put into an airtight tin or bottle and keep near the stove. Use sparingly in soups and stews.

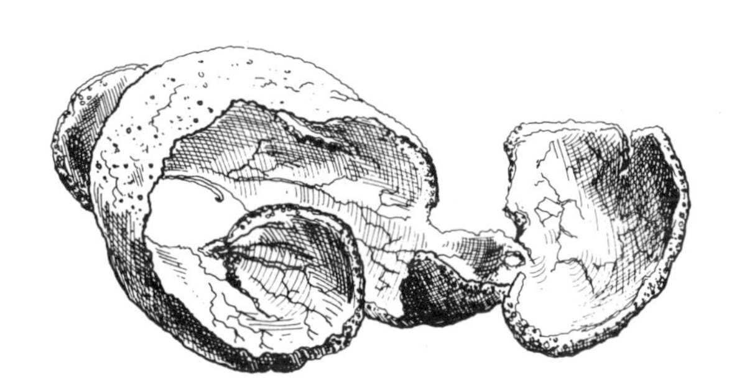

ROSEMARY OR ◆ LAVENDER SUGAR ◆

sprigs of fresh rosemary or lavender
caster sugar

Clean and dry the rosemary or lavender. Put into a screw-topped jar and fill with sugar. Shake well and leave for 24 hours. Shake again and leave for a week before use. Use the sugar in milk puddings of all kinds, or sprinkle on cakes or biscuits.

◆ ORANGE POWDER ◆

Remove as much pith from orange rind as possible. Wash the rind and dry slowly in a cool oven. When crisp, break in pieces and roll very finely or crush in an electric blender. Sieve and store in a jar or tin in a dry place. This is a strong flavouring which can be used for cakes and puddings, or mixed into Herb Seasoning.

◆ HERB AND FLOWER TEAS ◆

Herb and flower teas or tisanes made from leaves and flowers have a soothing effect (often promoting sleep) and can calm the digestion.

Elderflower Tea. Gather elderflowers and a few leaves on a dry summer day and spread out in the sun. When dry, remove from stems and store in paper bags or bottles. Use to soothe sore throats by infusing in boiling water (a small handful of mixed flowers and leaves in 600ml/1 pint water). Leave for 6 hours and sweeten with honey if liked.

Limeflower Tea. Gather the limeflowers on a sunny day, in late June or early July, dry in the sun or a warm place and store in bottles in the dark. Infuse for a few minutes in boiling water and sweeten with honey if liked. This is a very soothing tea and is an aid to digestion, so is often taken after an evening meal.

Primrose or Cowslip Tea. Free cowslips or primrose flowers from stalk or green, and dry in the sun or a warm place such as an airing cupboard, then store in jars in a dark place. Use a spoonful of the flowers in half a glass of boiling water to soothe and aid sleep.

Herb Teas. Mint, parsley, rosemary and sage are all good herb teas.

Mint serves as a digestive; parsley is supposed to be helpful for slight attacks of rheumatism; rosemary is a good pick-me-up and soothes colds, and sage soothes sore throats. Allow them to infuse in boiling water for 30 minutes and sweeten with honey.

• USING DRIED PEAS AND BEANS •

Pulses, or dried peas and beans, are full of nourishment, being high in protein, calcium and vitamins, and they can often be served as a substitute for more expensive protein foods. The wide variety of beans which can be home-grown and dried provide numerous colours, flavours and shapes which can give a new look to soups, casseroles and salads.

Keep home-dried beans dry in a tightly sealed, moisture-proof jar or other container, and keep each type separate. If kept in a cool place at a constant temperature, pulses will keep well for a year. Soak pulses in cold water before using them. They can be left overnight, but 3–6 hours is enough to soften them if they have not been stored too long. Beans will generally need simmering for 1–2 hours after soaking, and they should never be salted during cooking or they will become hard. After draining, they can be well seasoned and served with a dressing of butter, or tossed in an oil-and-vinegar dressing to serve cold.

To speed up the preparation of beans and peas without soaking, try covering them with water, bringing to the boil and boiling for 2 minutes. Take off the heat, cover and leave to stand for 1 hour. Bring back to the boil and simmer for an hour until tender but not broken. This initial preparation is not necessary when pulses are to be used in soup.

Beans and pulses may be prepared in a pressure cooker following manufacturers' instructions. Care should be taken to see that they are not overcooked and mushy.

Alternative Cooking Methods. Beans and pulses may be prepared in a slow-cooker, but red kidney beans must be boiled hard for 10 minutes before putting into the cooker to eliminate dangerous toxins.

There is little advantage in using a microwave oven for cooking beans and pulses as they will take just as long to cook as by normal methods.

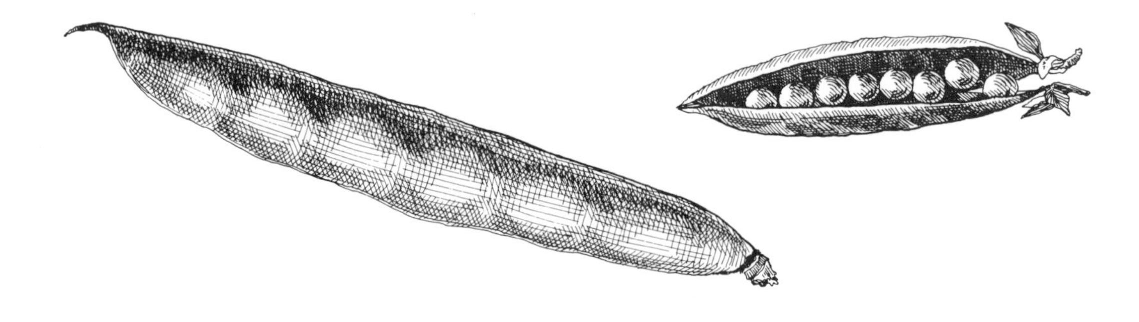

• DRIED PEA • SOUP

500g/1lb dried green peas

2 medium onions

2 carrots

1 small turnip

125g/4oz swede

1 bacon bone

salt

pepper

2 litres/4 pints water

Soak the peas overnight. Drain the peas and put them into a pan with chopped vegetables, the bacon bone, a little salt, pepper and fresh water. Bring to the boil, cover and simmer for about 2 hours until the peas are soft. Take out the bacon bone. Put the liquid and vegetables through a sieve or blend in a liquidiser until smooth. Reheat and add more seasoning if necessary. This soup is good with a garnish of crumbled crisp bacon, or fried bread cubes.

• BEEF & BEAN • CASSEROLE

125g/4oz dried haricot beans

1 large onion

2 tbsp oil

500g/1lb shin beef

300ml/½ pint tomato juice

600ml/1 pint water

50g/2oz concentrated tomato purée

salt

pepper

Soak the beans overnight and then drain well. Chop the onion and cook in the oil until soft and golden. Cut the beef into cubes and add to the onions, frying until sealed on all sides. Put the beef and onions into a casserole with the beans. Mix the tomato juice, water and tomato purée together and pour over the meat. Cover and bake at 170°C/325°F/gas mark 3 for 2 hours. Reduce heat to 150°C/300°F/gas mark 2 for 1 hour. Add a little more water or tomato juice during cooking if the dish becomes dry. Adjust seasoning before serving.

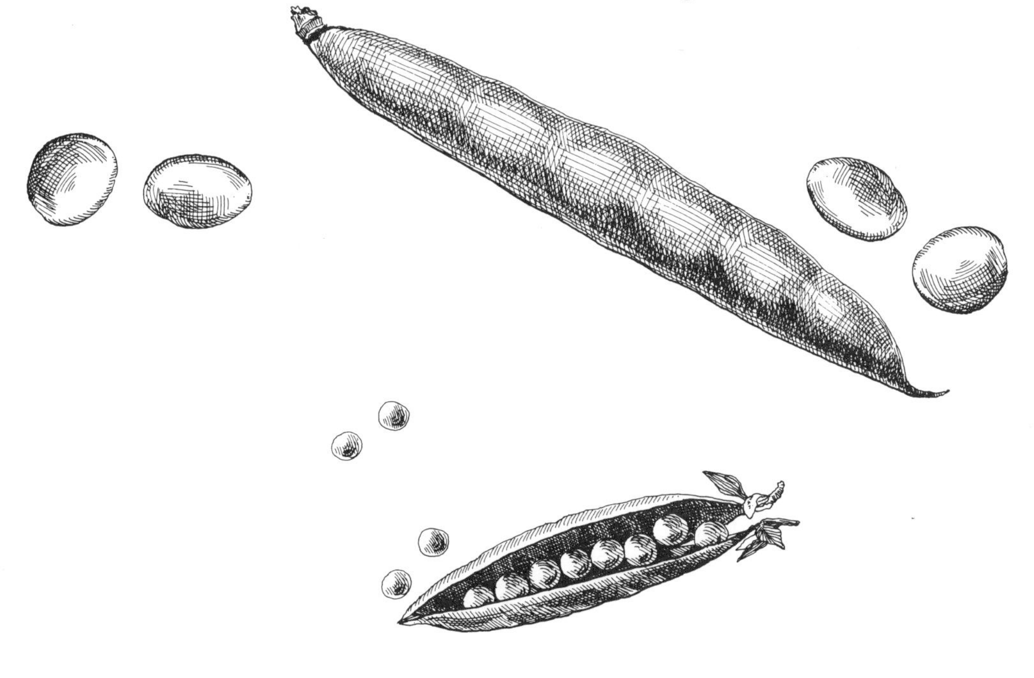

• BEANS •
IN CHEESE SAUCE

250g/8oz dried beans

2 celery sticks

1 medium onion

1 bay leaf

pinch bicarbonate of soda

40g/1½oz butter

40g/1½oz plain flour

450ml/¾ pint milk

50g/2oz grated Cheddar cheese

salt

pepper

pinch ground nutmeg

Use any dried beans for this dish, and soak them overnight. Drain; put into a pan with fresh water, chopped celery, onion, bay leaf and a pinch of soda. Bring to the boil and simmer for 1 hour. Drain and remove the bay leaf. Melt the butter and work in the flour. Cook for 2 minutes, then gradually stir in the milk and continue cooking and stirring until the sauce is smooth. Take off the heat and stir in the cheese. Add salt, pepper and nutmeg, and mix with the beans and vegetables. Put into a shallow oven-proof serving dish. Grill under a medium grill until the top is golden and bubbling. This dish is very good served with grilled tomatoes, or a tomato salad.

• HARICOT BEAN •
SALAD

250g/8oz dried haricot beans

2 medium onions

150ml/¼ pint olive oil

2 garlic cloves

1 bay leaf

pinch thyme

2 tsp concentrated tomato purée

salt

pepper

juice of 1 lemon

2 tbsp chopped parsley

Soak the beans overnight and drain. Chop 1 onion and cook in the oil until soft and golden. Add the beans, crushed garlic, herbs and tomato purée and cook gently for 10 minutes. Add enough boiling water just to cover the beans and simmer for 2 hours until the beans are tender but not broken and the liquid has been absorbed. Season well with salt, pepper and lemon juice and cool completely. Serve sprinkled with chopped parsley and topped with the remaining onion cut into thin rings.

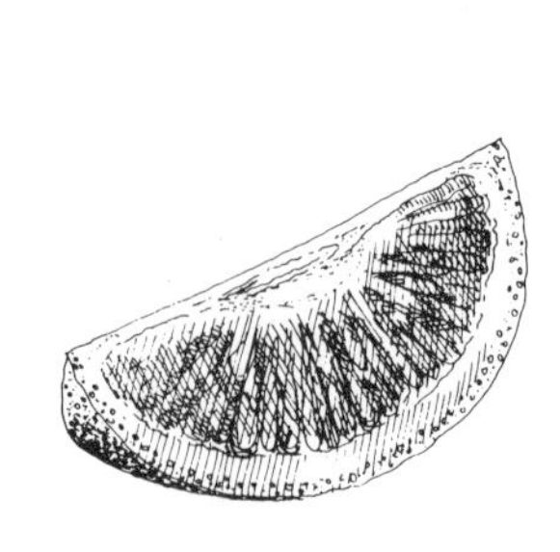

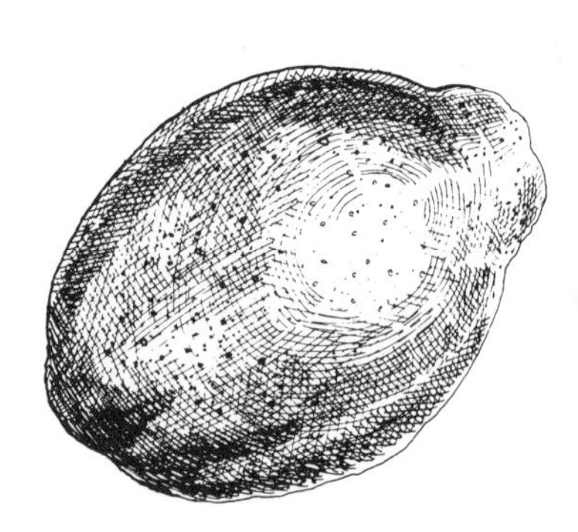

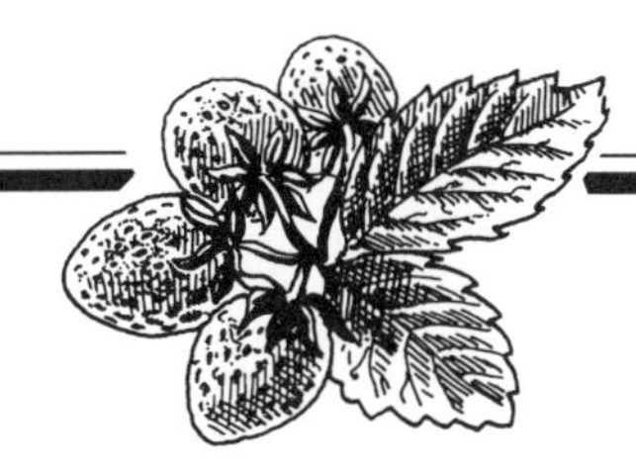

CLAMPING & CELLARING

If space is available, it is easy and cheap to store many kinds of fruit and vegetables by storing in outdoor clamps, or in cellars or sheds.

• VEGETABLES •

Green leafy vegetables should never be stored in this way as they quickly lose quality and vitamins. Peas and beans may be stored by drying (see the previous chapter). Jerusalem artichokes, celery, leeks and parsnips can be left in the ground until needed, but if there is likely to be snow or heavy frost, supplies for a few meals may be lifted, and the rest protected with a thick layer of straw.

Marrows, Pumpkins and Squashes. Cut these before autumn frost, and leave 3.75–5cm/1½–2in stalks. If possible, hang in nets. Ask a greengrocer for these, as many vegetables are delivered to him in nets which can easily be reshaped for home use. Small specimens may be put on shelves, but should be turned occasionally to prevent bruising or the growth of mould.

Onions, Shallots and Garlic. Only store well-dried bulbs from which the leaves have withered at the base. Small onions, or those not likely to keep long can be used up in pickles and chutneys. If the weather is wet, spread out the bulbs in a single layer in a shed or dry empty room until dry. Large onions may be tied together round the dry stems and hung in small bunches. They may also be hung in nets, or spread out in boxes in a single layer, but it is very important to keep them well ventilated or they will start to sprout. It is also possible to plait them together in long strings, or put them into an old clean nylon stocking, knotting the top when it is full.

Garlic should be dried and rubbed clean, and may be kept in the kitchen in a small wire basket, or plaited into a long string. Shallots are best netted or kept in wire baskets in the kitchen where they are easily accessible, but they should be kept away from dampness, excessive steam or warmth.

Root vegetables. Beetroot, carrots, turnips and swedes can all be stored in boxes or clamps (see Potatoes) well protected from frost and from mice. Long-rooted beetroot store best. Lift the vegetables in dry conditions, and rub off excess soil without damaging the roots. Twist off green tops and be sure all vegetables are sound.

Pack into boxes in layers with dry sand or wood ash in between to prevent shrivelling or frost damage and store in a dry shed or cellar. If the store is very dry, damp down the floor occasionally to help keep the vegetables in good condition. Complete darkness is essential for keeping these roots well.

If there is no shed, and no cellar in the house, the vegetables may be placed in boxes on boards standing on rafters in the roof, with the boxes covered with old sacking or several layers of crumpled newspapers. Cellars should be well ventilated, and if the floor is at all damp, it is wise to raise boxes on bricks.

Potatoes. There may well be a lot of potatoes to be stored in years when the crop is heavy, and this can present storage problems. All potatoes should be stored by the end of October, and the most common storing places are sheds, cellars and clamps. Whatever place is chosen must be dark and frost-proof as potatoes cannot stand frost and turn watery, becoming unfit for use. The storage place must also be dry and cool.

Potatoes should be lifted on a fine dry day, excess soil carefully brushed off, and the potatoes put into baskets as they are dug. They should not be left too long in the light as this harms potatoes, but may be left on the ground for an hour or two to dry any damp earth which may be on them.

In a cellar or shed, potatoes keep best spread out in heaps on the floor with long dry straw at the bottom, and a covering of dry straw at least 30cm/12in thick on the top and sides. Dry bracken may be used instead of straw. Any potatoes which are damaged in lifting should be left for immediate use.

Potatoes, as well as other root vegetables, may be safely stored outside in the open in mounds known as 'clamps', 'pies', or 'hogs'. The bottom of the clamp should be firm and level with ends facing north and south, so that the clamp may be opened at the south end in winter with the least likelihood of letting in frost. Use a well-drained site, and if the land is heavy, keep the clamp on ground level or slightly raised. On light soil, a shallow excavation of about 23cm/9in may be dug. A clamp is usually about 1m/3½ft wide and should be long enough to take all the potatoes.

Pile the potatoes in a ridge along the site and cover with dry straw or bracken a few inches thick. Bank this up with a layer of soil at least 30cm/12in thick, beat down flat with the back of a spade, leaving a little straw showing at the top for ventilation. Cover the sides and ends of this clamp with a layer of long straw about 10cm/4in thick, pressing the lower ends of the straw into the ground to prevent the entry of frost.

Place a layer of straw on the 'roof' of the clamp so that it overhangs the sides and rain will run off. Cover this straw with a layer of 7.5cm/3in deep soil.

When winter weather begins, dig a drainage trench about 30cm/12in wide and 15cm/6in deep all round the clamp and pile the surplus earth on the clamp. Cut away a short trench from the drainage trench so that water runs away. The layer of soil on top of the clamp should be about 15–20cm/6–8in deep to prevent frost entering. Clamping may seem to be a lot of trouble, but it does save on shed or cellar space and can preserve a large potato crop which can be bulky.

• FRUIT •

Apples and pears can be kept through to the spring without any need to bottle or freeze them. In general, early-maturing varieties will not keep long and should be used quickly. Late-maturing varieties will store for a long time however, and some of the best are:

Eating apples. Blenheim Orange, Cox's Orange Pippin, Laxton's Superb, Ribston Pippin and Sturmer Pippin.

Cooking apples. Annie Elizabeth, Bramley's Seedling, Edward VII, Lane's Prince Albert, Newton Wonder.

Pears. Catillac, Glous Morceau, Joesephine de Malines, Winter Nelis.

Fruit should be picked only when fully matured and when the stalk parts easily from the branch if the fruit is lifted gently. Leave very late varieties such as Sturmer Pippin on the tree until storms or frosts threaten. Store only sound fruit without bruises, scabs or cracks, or missing stalks, and handle the fruit carefully and individually, without dropping, throwing or tipping. Allow fruit to cool and sweat in a cool airy place overnight before storing.

Use a dark, cool and slightly moist store where it is possible to keep an even temperature and fairly constant humidity as this will prevent shrivelling and keep down disease. A temperature of 1–4°C/35–40°F is suitable for apples and slightly higher for pears. A shed, cellar or any outbuilding can be used, preferably with an earth, brick or concrete floor, and with good ventilation, and the store should be as dark as possible. It is most important that the store should be protected from rats or mice.

Keep special fruit wrapped separately in oiled papers, tissue or newspaper, as this will slow down ripening and prevent the spread of disease. Examine occasionally and throw away fruit which has spoiled. Pears ripen very quickly and soon spoil. If the store is not dark, cover with straw or sheets of paper.

The best storage for fruit is in single layers in trays, racks or boxes, or on the floor. Boxes may be packed one above the other if air circulates freely, and it is worth saving wooden boxes in which tomatoes and soft fruit are often packed as these have projecting corners for easy stacking.

Small fruit which will be mainly used for cooking can be stacked in layers with straw between (or individually wrapped), but the stack should only be 60cm/2ft high.

If a store becomes too warm when the weather is mild, ventilate it at night, but not when there is risk of frost. An upstairs room or other dry place is not suitable for storage as fruit quickly shrivels. If there is no covered space, boxes may be stacked in the open, raised on bricks or boards above ground level to prevent the entry of damp, and with adequate ventilation. Wire netting may be necessary to protect fruit from mice, and a good covering of straw should be used to protect from frost or rain. This method is really only for short-term storage, and pears should not be stored outside.

• NUTS •

Many nuts store well and are worth saving for use in recipes. All nuts should be left to mature before gathering, and should then be stored in shells to keep fresh right through the winter.

Almonds. There are two types of almonds – bitter and sweet. Most almonds grown in Britain are bitter almonds which may be used in small quantities for flavouring, but should not be eaten for dessert. The sweet almond is the one commonly used for dessert and for cooking. Remove any husk and fibre from the shells with a dry nail brush. Spread out the nuts in a single layer at room temperature with a current of air until the shells are dry. Store in a cool place. If shells are difficult to crack after storage, put the nuts in a warm oven for a few minutes and this will make them easier to crack.

Cobnuts. These are best gathered when the husks begin to dry in September or early October. Spread them out in a single layer in an airy place and turn them often to prevent mildew. Do not keep cobnuts beyond the end of December.

Sweet Chestnuts. Remove the husks and wipe the nuts. Store in sacks or boxes in a cool dry place.

Walnuts. Gather these as soon as they fall and clean them as soon as possible as the green outer husk quickly becomes black and difficult to remove. Take off the husks and free the walnut shells from any trace of fibre which will encourage mould. Scrub with a dry nail brush, but if the fibres are obstinate, scrub with a little water, but do not leave the nuts soaking or they will begin to crack along the seam. Dry the nuts in a single layer at room temperature in a current of air.

SWEET PRESERVES

All kinds of delicious sweet preserves can be made from fruit and some vegetables. They are useful for breakfast and teatime, but also for making tarts, and filling cakes, and for providing sauces for puddings and ices. Equipment can be very simple, and preparation is easy once the basic principles which govern the setting of preserves are understood. When making jams, try a number of different recipes for each fruit. It is worth mixing different fruits for variety, or making a number of different jams rather than one huge batch of a single recipe.

• EQUIPMENT •

Use a large preserving pan in which preserves can be boiled rapidly without boiling over. The pan should be wide to allow rapid evaporation of liquid which is essential to good setting. Use an aluminium or copper pan (chipped enamel is dangerous, while zinc or iron spoil colour and flavour). Copper will keep green fruits green, but can spoil the colour of red fruit.

Additional useful items are a metal or heat-proof glass jug for pouring jam into jars cleanly and easily; a metal jam funnel; a long-handled wooden spoon which prevents hot jam splashing on the hands, and accurate scales. A sugar thermometer takes the guesswork out of jam-

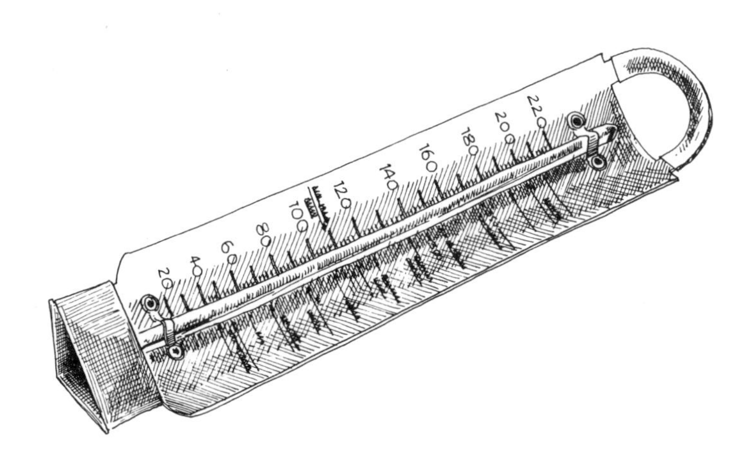

making as it records the temperature at which preserves set, but it is not essential. A jelly bag or clean piece of cloth will be necessary for jelly-making, and some pieces of muslin for tying up spices or fruit pips which contain pectin.

Jam jars may be saved from year to year, and screw-top honey jars, coffee jars and preserving jars are very useful. Waxed discs, transparent covers and labels can be bought in packets, or tight-fitting plastic covers can be used, or twist tops which do not need a waxed disc.

• INGREDIENTS •

• Fruit

Fresh sound fruit should be used, which is not wet or mushy, and it should be slightly under-ripe, as very ripe fruit has a reduced sugar content and will affect the setting and keeping quality of preserves. Fruit should be sorted, trimmed, washed gently and patted dry, but hot water should never be used, and low-pectin fruit should be washed as little as possible. Only peel, cut or stone fruit just before cooking, or the quality will deteriorate.

Acid is added to some fruit during cooking to extract pectin, improve colour and prevent crystallization – it should be added to any fruit with a low acid content, and to any vegetable jam. Acid may be in the form of lemon juice, citric or tartaric acid, redcurrant or gooseberry juice. To 2kg/4lb fruit, allow 2 tbsp lemon juice *or* ½ tsp citric or tartaric acid *or* 150ml/¼ pint redcurrant or gooseberry juice.

• Pectin

The setting of jam depends on its pectin content, and some types of fruit have a higher pectin content than others, so that they are good 'mixers' with other fruits. Cooking apples, blackcurrants, damsons, gooseberries, plums, quinces and redcurrants are all high in pectin, and preserves made with them always set well. Fresh apricots, early blackberries, greengages and loganberries have a lower pectin content.

Fruit and vegetables which are low in pectin include late blackberries, cherries, elderberries, marrows, medlars, pears, rhubarb, strawberries and tomatoes, and these need mixing with high-pectin fruit, or juice made from such fruit, while acid will help to release the pectin.

• Sugar

Preserving sugar dissolves quickly and is ideal for jam-making as it gives a very clear result, but it is expensive and not always easy to obtain. Cube or granulated sugar is perfectly satisfactory. Sugar must be stirred

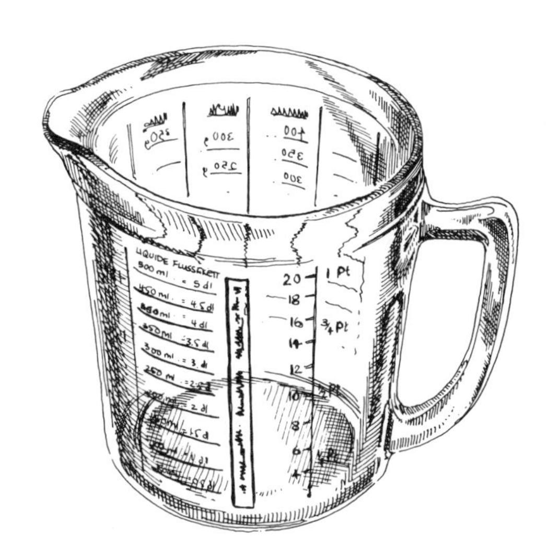

carefully into the cooked fruit until dissolved before hard boiling, or crystals will remain and burn on the bottom of the pan, and will also affect the smoothness of the jam. Sugar may be warmed slightly before adding to the fruit to speed up the dissolving process.

◆ Making the preserves

Whether jam, jelly, marmalade or curd is being made, it is important to be methodical. Assemble all equipment and ingredients before cooking, and follow recipes carefully for perfect results. Cook fruit slowly to extract pectin, soften skins and keep a good colour, and then boil quickly without stirring to give a high yield with good colour and flavour.

The preserve must be tested for setting, skimmed and poured into hot clear jars, filled to the brim. Waxed circles should be put straight on to the hot jam, and then the covers put on when the preserve is hot or completely cold. Jars must be carefully cleaned, labelled and stored in a cool dry place.

◆ Setting tests

A keeping jam should have 60 per cent added sugar content or three parts sugar to five parts jam. Some jams are ready for setting after only 5 minutes' boiling; others take 10–15 minutes, and a few take longer. A setting test should be made early, as some fruits lose their setting qualities if jam is boiled too long and if this happens the jam will never set. When the jam reaches setting point, remove it from heat at once.

Temperature test. Place a sugar thermometer in hot water. Stir the jam and submerge the thermometer bulb completely in the jam. When it registers 110°C/220°–221°F, the jam is cooked.

Weight test. Weigh the pan and spoon before cooking begins. When the jam weighs 5kg/10lb for every 3kg/6lb of sugar used, the jam is ready. To work out the final correct weight, multiply the quantity of sugar used by ten and divide by six.

Flake test. Dip a clean wooden spoon in the boiling jam. Let the cooling jam drop from the spoon, and if the drops run together and form a flake, the jam is ready.

Plate test. Put a little jam on an old plate or saucer and leave it to become cold. If the jam forms a skin and wrinkles when pushed with a spoon or finger it is ready. The jam pan should be taken off the heat while the test jam is cooling.

• Preserving faults

Mouldy jam results from damp fruit, insufficient boiling, poor storage, badly-filled or covered jars.

Crystallized jam indicates too much sugar in proportion to fruit or over-cooking jam to stiffen it when too little sugar has been used. Over-cooking and poor stirring resulting in undissolved sugar can also cause the problem.

Fermenting, 'winey' jam results from over-ripe fruit, insufficient sugar or boiling, poor covering and bad storage.

Hard, dry jam results from over-boiling or bad covering when jam is stored in a warm place. Plastic or screwtops will help to prevent this problem.

Syrupy jam or jelly results from lack of pectin, from insufficient boiling or over-boiling past setting point. Jelly can be affected if strained fruit juice is left too long before cooking.

Poor flavour results from over- or under-ripe fruit, too much sugar, too slow boiling or over-boiling.

Poor colour results from poor quality fruit, from a poor quality preserving pan, or from storing in a bright light. It can also arise if the fruit is not cooked slowly enough to soften it completely, if the jam is over-boiled, or boiled too slowly to setting point.

Cloudy jelly results from poorly strained juice through a bag which is too thin, or if pulp is forced through the bag instead of letting the jelly drip by itself.

◆ JAM ◆

There is an endless variety of jam to be made, particularly if fruits are combined. This means that some common fruit such as apples and rhubarb can be used as bulk ingredients combined with more special fruit such as raspberries to give more interesting jams in larger quantities.

Wash and drain fruit and take out stems, leaves and bruised or damaged pieces. Stone fruit may be cooked whole and the stones removed during cooking, or the stones may be removed beforehand. Apricot, cherry, plum and greengage jams can be flavoured with a few kernels from their stones. The fruit should be cooked slowly in water until completely tender, adding acid during this cooking if specified. When the fruit is soft and the contents of the pan reduced by about one half, the sugar should be stirred in over low heat until dissolved. After that, the jam must be boiled rapidly to setting point, and should be tested after 5 minutes. Some jams take longer, but no jam needs to boil longer than 20 minutes. After testing for setting, skim the jam and leave it to cool for a minute or two, before stirring so that fruit does not rise in the jars. Pour into hot clean jars, filling right to the brim, and put on waxed discs at once, pressing carefully to exclude air bubbles. Put on top covers and clean the jars before labelling and storing in a cool dry place.

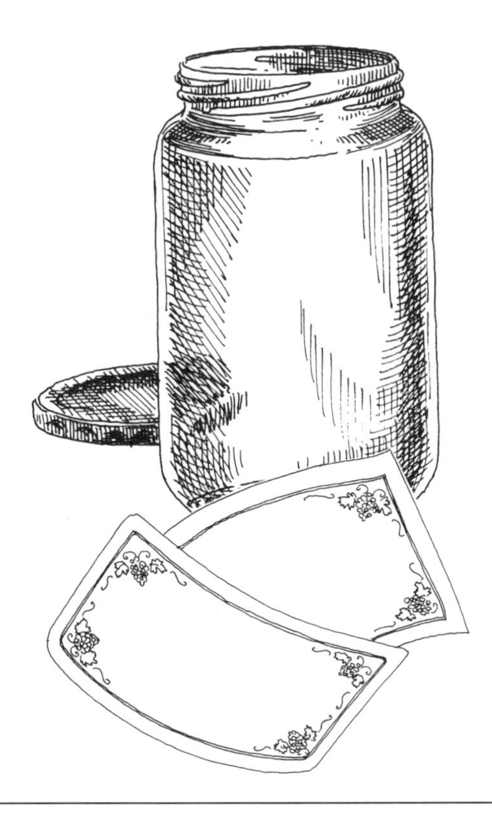

• APPLE JAM •

1.5kg/3lb cooking apples

600ml/1 pint water

2 tsp citric acid

6 cloves

sugar

Cut the apples into slices without peeling or coring. Put into the pan with water, acid and cloves and simmer to a pulp. Take out the cloves and sieve the apples. Weigh the pulp and allow 350g/12oz sugar to each 500g/1lb apple pulp. Stir in the sugar over low heat until dissolved, and then boil hard to setting point. Pour into hot jars and cover.

• APPLE GINGER •

1.5kg/3lb cooking apples

600ml/1 pint water

1 tsp ground ginger

2 lemons

1.5kg/3lb sugar

125g/4oz crystallized ginger

Peel and core the apples and put the peel and cores into a piece of muslin. Tie into a bag-shape and suspend them in the pan. Cut the apples into pieces and put into the pan with the water, ground ginger, grated lemon rind and juice. Simmer over low heat until the fruit is soft. Take out the bag of peel and cores, and squeeze the juice into the apples. Stir in the sugar and chopped ginger over low heat until the sugar has dissolved. Boil hard to setting point, pour into hot jars and cover.

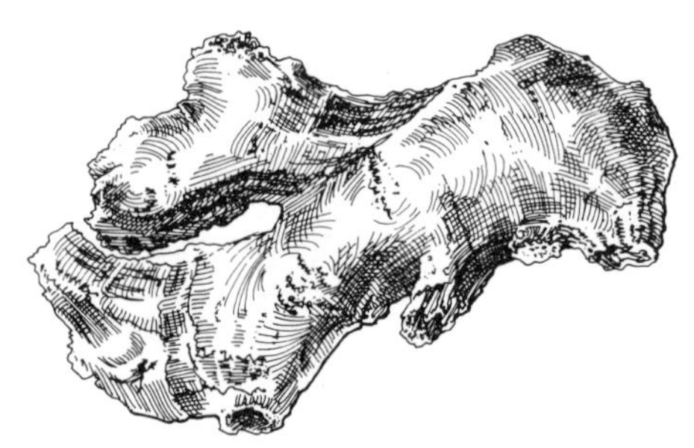

FRESH • APRICOT JAM •

3kg/6lb fresh apricots

450ml/¾ pint water

1 lemon, juice

3kg/6lb sugar

Wash the apricots and cut them in half. Take out the stones. Put the fruit, water and juice into a pan and simmer until the fruit is soft. Meanwhile, take the kernels from a few of the stones, blanch them in boiling water, and split them in half. Add to the fruit while it is cooking. Stir in the sugar over low heat until it has dissolved. Boil hard to setting point, pour into hot jars and cover.

DRIED • APRICOT JAM •

500g/1lb dried apricots

1.5 litres/3 pints water

1 lemon, juice

1.5kg/3lb sugar

75g/3oz blanched almonds

Cut the apricots in small pieces with scissors and leave them to soak in the water for 24 hours. Put the fruit and water into a pan and cook gently for 30 minutes. Add the lemon juice, sugar and halved almonds, and stir over low heat until the sugar has dissolved. Boil hard to setting point, pour into hot jars and cover.

• AUTUMN JAM •

1kg/2lb cooking apples

1kg/2lb ripe pears

1kg/2lb plums

25g/1oz root ginger

2.5kg/5lb sugar

Peel and core the apples and pears and put into a pan. Skin the plums, cut in half and remove the stones. Add to the other fruit in the pan. Bruise the ginger and tie it into a piece of muslin, suspending the bag in the pan. Simmer gently until the fruit is soft but not broken, adding a little water if necessary to prevent burning, although the fruit will yield plenty of juice. Stir in the sugar over low heat until dissolved. Boil hard to setting point. Remove the ginger. Pour into hot jars and cover.

• BLACKBERRY JAM •

3kg/6lb blackberries

150ml/¼ pint water

4 tsp lemon juice

3kg/6lb sugar

Wash the blackberries and remove any stems and unripe berries. Put the fruit into a pan with the water and lemon juice. Simmer until the fruit is very soft. Stir in the sugar over low heat until dissolved. Boil hard to setting point, pour into hot jars and cover.

BLACKBERRY & • APPLE JAM •

2kg/4lb blackberries

300ml/½ pint water

1kg/2lb cooking apples

3kg/6lb sugar

Wash the blackberries and remove any stems and unripe berries. Put the berries into a pan with half the water and simmer until tender. Peel, core and slice the apples, and cook in the remaining water until soft but not broken. Combine the two fruits and liquid in one pan, and stir in the sugar over low heat until dissolved. Boil hard to setting point, pour into hot jars and cover.

BLACKBERRY & • ELDERBERRY JAM •

1kg/2lb elderberries

1kg/2lb blackberries

1.5kg/3lb sugar

Wash the elderberries and strip them from their stalks. Wash the blackberries and remove any stems and unripe berries. Mix the two fruits together and simmer for about 20 minutes, crushing the berries with a wooden spoon to release the juices. When the fruit is tender, stir in the sugar over low heat until dissolved. Boil hard to setting point, pour into hot jars and cover.

BLACKBERRY & • RHUBARB JAM •

2kg/4lb blackberries
1kg/2lb rhubarb
450ml/¾ pint water
sugar

This is a good jam to make from the last of the rhubarb crop which will be finishing when the blackberries come into season. Wash the berries and remove any stems and unripe berries. Simmer in the water until tender and sieve to remove all seeds. Wash and cut up the rhubarb and put into the blackberry pulp. Simmer until soft and then weigh the fruit. Allow 500g/1lb sugar to each 500g/1lb fruit. Stir in the sugar over low heat until dissolved. Boil hard to setting point, pour into hot jars and cover.

• BLACKCURRANT • JAM

2kg/4lb blackcurrants
1.5 litres/3 pints water
3kg/6lb sugar

Wash the fruit and remove the stems. Put into a pan with the water and simmer gently until the fruit is soft, stirring often to prevent burning. Stir in the sugar over low heat until dissolved. Boil hard to setting point, pour into hot jars and cover.

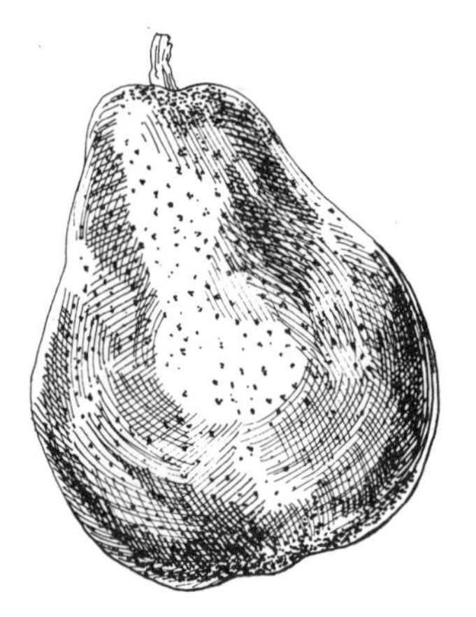

• CHERRY JAM •

2kg/4lb black cherries
8g/¼oz citric or tartaric acid
1.5kg/3lb sugar

Take the stones out of the cherries. Crack 24 stones and remove the kernels. Put the cherries into a pan with the kernels and acid and simmer until very soft, stirring often to prevent burning. Stir in the sugar over low heat until dissolved. Boil hard to setting point, pour into hot jars and cover. Do not expect a strong set from cherry jam.

• DAMSON JAM •

1.5kg/3lb damsons
450ml/¾ pint water
1.5kg/3lb sugar

Wash the damsons and simmer them in the water until soft, removing any stones as they rise in the pan. Stir in the sugar over low heat until dissolved. Boil hard to setting point, pour into hot jars and cover.

DAMSON & • PEAR JAM •

1kg/2lb damsons
1kg/2lb ripe pears
300ml/½ pint water
2kg/4lb sugar

Wash the damsons and put them into a pan. Peel and core the pears and add them to the damsons with the water. Simmer until soft, removing damson stones as they rise to the surface. Stir in the sugar over low heat until dissolved. Boil hard to setting point, pour into hot jars and cover.

• FRESH FIG JAM •

1kg/2lb fresh figs
500g/1lb cooking apples
6 lemons, juice and grated rind 2 lemons
1kg/2lb sugar

This is a useful recipe for figs which grow in a cool climate where they form well but seldom ripen enough to be pleasant for dessert eating. The jam has a delicate and unusual flavour. Wash the figs, peel them, and blanch in boiling water for 1 minute. Drain well, rinse in cold water, and cut in thin slices. Peel, core and slice the apples and mix with the figs, lemon juice and rind in a pan. Cover and simmer until the fruit is tender, stirring often. Stir in the sugar over low heat until dissolved. Boil hard for 15 minutes, pour into hot jars and cover. If a spiced flavour is liked, put a small piece of cinnamon stick, a piece of bruised root ginger and 2 whole cloves in a piece of muslin and suspend in the pan while the jam is cooking (remove before bottling).

• GOOSEBERRY • JAM

3kg/6lb gooseberries
1 litre/2 pints water
3kg/6lb sugar

The best gooseberries for jam are slightly under-ripe. Wash the fruit, top and tail, and put into a pan with the water. Simmer for about 30 minutes until soft, mashing the fruit and stirring well. Stir in the sugar over low heat until dissolved. Boil hard to setting point, pour into hot jars and cover. Gooseberries which remain green when ripe will yield a green jam if the fruit and sugar are cooked quickly together – prolonged boiling results in a red jam. Fully-ripe or dessert gooseberries yield a lightly setting pink jam. For a delicate muscat flavour, tie a few elderflower heads in muslin and suspend them in the pan while cooking the fruit in water.

• GREENGAGE • JAM

3kg/6lb greengages
600ml/1 pint water
3kg/6lb sugar

Wash the greengages and cut them in half. Take out the stones and crack a few of them. Remove the kernels, blanch in boiling water and cut the kernels in half. Put the fruit and kernels into a pan with the water and simmer until the fruit is soft. Stir in the sugar over low heat until dissolved. Boil hard to setting point, pour into hot jars and cover.

• JAPONICA • JAM

2kg/4lb japonica fruit
3 litres/6 pints water
2 tsp ground cloves
sugar

The ornamental japonica yields a fruit which is a type of quince and has a similar flavour. Wash the fruit, but do not peel or core. Slice the fruit and simmer in the water until tender. Put through a sieve and weigh the pulp. Add an equal weight of sugar and the cloves. Stir over low heat until the sugar had dissolved. Boil hard to setting point, pour into hot jars and cover.

• MARROW GINGER •

3kg/6lb prepared marrow

4 lemons

75g/3oz root ginger

3kg/6lb sugar

A very large marrow will be needed to yield enough prepared flesh, but the quantities may be halved if preferred. Peel the marrow, remove seeds and pith and weigh the flesh. Cut into neat cubes and steam until just tender. Put into a bowl with the grated rind and juice of the lemons. Bruise the ginger and tie in a piece of muslin to suspend in the bowl. Stir in the sugar. Leave in a cool place for 24 hours. Put all the ingredients into a pan and stir over low heat until the sugar has dissolved. Cook gently until the narrow cubes are transparent and the syrup is thick. Discard the ginger. Pour into hot jars.

• MEDLAR JAM •

2kg/4lb medlars

sugar

vanilla pod

Medlars are old-fashioned fruit shaped like large rose-hips and the colour of unripe russet apples. They are very hard and should be gathered when fully formed, then kept for some weeks until soft or 'bletted', when the flesh will be brown. For the jam, scrape the pulp from very ripe medlars and cook very gently until soft, adding very little water if necessary so that they do not burn. Sieve and weigh the pulp. Add 350g/12oz sugar for each 500g/1lb fruit pulp. Stir over low heat until the sugar has dissolved. Put in the vanilla pod, and boil hard to setting point, stirring well. Take out the vanilla pod, pour into hot jars and cover.

• MULBERRY JAM •

1kg/2lb mulberries

1 lemon

750g/1½lb sugar

Rinse the mulberries gently so that no juice is lost. Put into a pan and simmer until soft. Add the grated lemon rind and lemon juice and stir in the sugar over low heat until dissolved. Boil hard to setting point, pour into hot jars and cover.

• PEACH JAM •

1kg/2lb ripe peaches

1 lemon, juice

750g/1½lb sugar

2 tsp rosewater

2 tsp orange flower water

Skin the peaches by dipping them into boiling water and then cold water. Cut the peaches in quarters, take out the stones, and put the fruit into a pan with the lemon juice. Simmer gently until the fruit is soft, stirring well and adding a little water if necessary to prevent burning. Stir in the sugar over low heat until dissolved. Boil hard to setting point. Stir in the rosewater and orange flower water and simmer for 1 minute. Pour into hot jars and cover.

• PEAR GINGER •

2kg/4lb cooking pears

2kg/4lb sugar

2 tsp ground ginger

2 lemons

Peel and core the pears. Chop the flesh finely, or mince the pears. Put into a bowl with the sugar and leave overnight. Put into a pan with the ginger, grated rind and juice of the lemons. Simmer until the fruit is tender, stirring well. Boil hard to setting point, pour into hot jars and cover.

• PLUM JAM •

3kg/6lb plums

600ml/1 pint water

3kg/6lb sugar

Wash the plums, cut them in half and take out the stones. Remove the kernels from a few of the stones, blanch them in boiling water, and cut them in half. Put fruit, water and kernels in a pan and simmer until the fruit is very soft. Stir in the sugar over low heat until dissolved. Boil hard to setting point, pour into hot jars and cover.

PLUM & • APPLE JAM •

1kg/2lb plums

1kg/2lb cooking apples

900ml/1½ pints water

1.5kg/3lb sugar

Wash the plums, cut them in half and take out the stones. Peel and core the apples, and cut them in slices. Put the fruit and water into a pan and simmer until soft. Stir in the sugar over low heat until dissolved. Boil hard to setting point, pour into hot jars and cover.

• QUINCE JAM •

2kg/4lb quinces

water

sugar

Peel and core the quinces and cut the flesh into small cubes. Put into a pan with just enough water to cook without burning. Simmer until the fruit is soft but unbroken. Weigh the fruit and liquid and allow 500g/1lb sugar to each 500g/1lb fruit. Stir in the sugar and leave overnight. Bring to the boil and then simmer for 20 minutes. Pour into hot jars and cover.

• RASPBERRY JAM •

25g/1oz butter

1.5kg/3lb raspberries

1.5kg/3lb sugar

Warm a pan and rub it with the butter. Put in the raspberries and heat very slowly until the juice runs. Warm the sugar in the oven. Add the warm sugar to the raspberries and heat over a low heat for 30 minutes. Pour into hot jars and cover. Raspberry jam made this way will keep a fresh red colour and taste like the fresh fruit.

RASPBERRY & • REDCURRANT JAM •

750g/1½lb raspberries

750g/1½lb redcurrants

600ml/1 pint water

1.5kg/3lb sugar

Wash the raspberries. Wash and string the redcurrants. Mix the fruit together in a pan with the water and simmer for 20 minutes. Stir in the sugar over low heat until dissolved. Boil hard to setting point, pour into hot jars and cover.

RASPBERRY & ◆ RHUBARB JAM ◆

1.5kg/3lb raspberries

1.5kg/3lb rhubarb

300ml/½ pint water

3kg/6lb sugar

Wash the raspberries. Wash the rhubarb and cut into chunks. Simmer the rhubarb in the water until soft. Add the raspberries and continue cooking until the raspberries are soft. Stir in the sugar over low heat until dissolved. Boil hard to setting point. Pour into hot jars and cover.

RHUBARB & ◆ FIG JAM ◆

1kg/2lb rhubarb

250g/8oz dried figs

1kg/2lb sugar

1 lemon, juice

Wash the rhubarb and cut into chunks. Cut the figs into small pieces. Mix the fruit, sugar and lemon juice in a bowl and leave to stand for 24 hours. Put into a pan and bring to the boil. Boil rapidly to setting point. Cool for 15 minutes and then stir well to distribute the figs. Pour into hot jars and cover.

◆ STRAWBERRY ◆ JAM

2kg/4lb strawberries

1 tsp citric or tartaric acid

1.75kg/3½lb sugar

Hull the strawberries and wash them. Put into a pan with the acid and simmer for 30 minutes until the fruit is soft. Stir in the sugar over low heat until dissolved. Boil hard to setting point. Cool for 15 minutes, stir well to distribute the fruit, pour into hot jars and cover.

◆ SUMMER JAM ◆

250g/8oz blackcurrants

250g/8oz redcurrants

250g/8oz raspberries

250g/8oz strawberries

1kg/2lb sugar

Wash all the fruit and remove stems of blackcurrants and redcurrants. Put the blackcurrants into a pan with very little water and simmer until tender. Add the remaining fruit and simmer for 10 minutes. Stir in the sugar over low heat until dissolved. Boil hard to setting point, pour into hot jars and cover.

◆ GREEN TOMATO ◆ JAM

1kg/2lb green tomatoes

1 sweet orange, rind

750g/1½lb sugar

Cut up the tomatoes and put into a pan. Shred the orange rind very finely and simmer in a little water until tender. Drain and add to the tomatoes. Simmer until the tomatoes are tender. Stir in the sugar over a low heat until dissolved. Boil hard to setting point. Pour into hot jars and cover.

♦ Freezer jam

Uncooked jam can be stored in the freezer for six months and is brightly coloured with the delicious smell of fresh fruit. Since the fruit is ripe and uncooked, it retains the maximum fresh fruit flavour. Ingredients must be measured carefully, and instructions followed closely. The jams contain a high proportion of sugar, and the yield for each 500g/1lb of fruit is high. The jams are mixed and left to set before freezing. After the first five hours at room temperature, the jams should be put into a refrigerator to finish setting before storage in the freezer. They are best packed in small quantities in rigid containers with tight-fitting lids, allowing 12mm/½in headspace.

Freezer jam should be stored at −18°/0°F or lower, and will keep for six months. Sometimes a white mould-like formation is noticeable when jam is removed from the freezer, but this is not harmful and will melt quickly at room temperature. Jams should be thawed for 1 hour before serving. After opening, they should be stored in the refrigerator and used up quickly. If uncooked jam is stiff, or if 'weeping' has occurred, it should be lightly stirred to soften and blend before serving.

FREEZER • APRICOT OR PEACH • JAM

750g/1½lb fresh apricots or peaches

1 tsp citric acid

1kg/2lb caster sugar

125ml/4fl. oz liquid pectin

Skin the apricots or peaches and remove the stones. Mash the fruit and stir in citric acid and sugar. Leave for 20 minutes, stirring occasionally, then add pectin and stir for 3 minutes. Put into small containers, cover and seal. Leave for 5 hours at room temperature, then put into refrigerator until jelled. This may take 24–48 hours. Store in freezer. Thaw for 1 hour at room temperature before serving.

FREEZER • BLACKBERRY JAM •

750g/1½lb blackberries

1.25kg/2¾lb caster sugar

125ml/4fl. oz liquid pectin

Small hard wild blackberries are difficult to mash without liquid and may be 'pippy', so this jam is better made with large cultivated blackberries. Mash the blackberries and stir in sugar. Leave for 20 minutes, stirring occasionally, then add pectin and stir for 3 minutes. Put into small containers, cover and seal. Leave for 5 hours at room temperature, then put into refrigerator until jelled. This may take 24–48 hours. Store in freezer. Thaw for 1 hour at room temperature before serving.

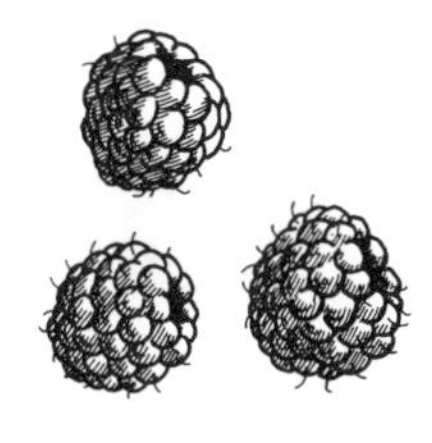

FREEZER
• CHERRY JAM •

750g/1½lb Morello cherries

1kg/2lb caster sugar

125ml/4fl. oz liquid pectin

Remove stones from cherries and put through coarse blade of a mincer. Stir with sugar and leave for 20 minutes, stirring occasionally. Add pectin and stir for 3 minutes. Put into small containers, cover and seal. Leave for 5 hours at room temperature, then put into refrigerator until jelled. This may take 24–48 hours. Store in freezer. Thaw for 1 hour at room temperature before serving.

FREEZER
• PLUM JAM •

750g/1½lb Victoria plums

1kg/2lb caster sugar

2 tsp lemon juice

125ml/4fl. oz liquid pectin

Remove stones from plums and stir in sugar and lemon juice. Leave for 20 minutes, stirring occasionally. Add pectin and stir for 3 minutes. Put into small containers, cover and seal. Leave for 5 hours at room temperature, then put into refrigerator until jelled. This may take 24–48 hours. Store in freezer. Thaw for 1 hour at room temperature before serving.

FREEZER
• RASPBERRY JAM •

750g/1½lb raspberries

1.5kg/3lb caster sugar

125ml/4fl. oz liquid pectin

Mash or sieve the raspberries and stir with the sugar. Leave for 20 minutes, stirring occasionally, then add pectin and stir for 3 minutes. Put into small containers, cover and seal. Leave for 5 hours at room temperature, then put into refrigerator until jelled. This may take 24–48 hours. Store in freezer. Thaw for 1 hour at room temperature before serving.

FREEZER
• STRAWBERRY JAM •

750g/1½lb strawberries

1kg/2lb caster sugar

125ml/4fl. oz liquid pectin

Mash or sieve the strawberries and stir with the sugar. Leave for 20 minutes, stirring occasionally, then add pectin and stir for 3 minutes. Put into small containers, cover and seal. Leave for 5 hours at room temperature, then put into refrigerator until jelled. This may take 24–48 hours. Store in freezer. Thaw for 1 hour at room temperature before serving.

• MARMALADE •

Citrus fruit marmalade is a delicious preserve easily made at home. While Seville oranges are only in season for a few weeks at the beginning of the year and the main marmalade-making time is usually in January, it is possible to make marmalade from sweet oranges, grapefruit, lemons and tangerines, and batches of preserve may be made throughout the year. It is particularly useful to be able to make extra marmalade in the autumn to see the household through until the New Year, and this is a good time to experiment with the other fruits. Since Seville oranges have such a limited season, it can be worth freezing them for later use. They need only be wiped and packed in 1kg/2lb bags. When needed for marmalade-making, they can be treated exactly the same as for fresh fruit.

To ensure a good set with home-made marmalade, the fruit or peel should be sliced, simmered in water until tender, and then boiled rapidly with sugar to setting point. A blender and a pressure cooker will help to speed up this initial cooking time considerably. If peel and sugar are boiled too long together, the result will be a thick syrup with hard chunks of peel. The fruit need not be soaked overnight, but when the peel is cooked it should be completely soft when tested with the fingers. The liquid in this first cooking must be evaporated and generally the contents of the pan will be reduced by about half.

It can be tiring cutting up fruit for marmalade, but an attachment is now available on some electric mixers which will make this much easier. If the fruit has a great deal of pith, the peel should be removed before slicing. If oranges are thin-skinned, it is possible to cut them in quarters lengthwise without peeling, and then the fruit and peel can be sliced through together in thin pieces. Thin-skinned lemons can also be cut this way, but grapefruit and thick-skinned lemons will need peeling first, and then the pith should be removed before cutting the flesh. Peel swells during cooking, so it is best to cut it thin. Pips, white pith and any trimmings should be put into a piece of muslin tied into a bag shape and suspended in the pan during the first cooking. Valuable pectin is contained in this mixture which helps the marmalade to set. The bag should be taken out and all liquid squeezed into the pan before the sugar is added.

Sugar should be stirred in over low heat so that it is completely dissolved before boiling. After that, it is essential to boil the fruit and sugar mixture hard to setting point, which will probably take about 15 minutes. The easiest test is to pour a little on to a cold plate, leave it for a minute to cool and then push the marmalade with a finger. If it wrinkles, the marmalade is ready. Remove from the heat and leave to stand for at least 5 minutes, then stir well to prevent peel rising to the top. Pour marmalade into clean, warm jars, filling right to the top, and cover with waxed paper circles. Cover at once or when completely cold.

• BLENDER CHUNKY • MARMALADE

1kg/2lb Seville oranges

1 lemon

2 litres/4 pints water

2kg/4lb sugar

Cut the fruit in quarters and take out the pips. Tie the pips in a piece of muslin and suspend in a pan. Put small quantities of the fruit with some of the water into an electric blender and blend on high speed until chopped. Pour into the pan and continue the process until all the fruit has been chopped. Add any remaining water. Simmer for 1 hour. Take out the bag of pips and squeeze the juice into the pan. Stir in the sugar over low heat until it has dissolved. Boil hard to setting point. Cool for 5 minutes, stir well, pour into hot jars and cover. This is a very quick marmalade to make, with an excellent set and flavour.

• GINGER • MARMALADE

5 Seville oranges

2.5 litres/5 pints water

1.5kg/3lb cooking apples

3kg/6lb sugar

250g/8oz crystallized ginger

15g/½oz ground ginger

Cut the oranges in half and squeeze out the juice. Cut up the flesh and shred the peel finely. Put the orange peel and flesh, juice and water into a pan. Tie the pips into a piece of muslin and suspend in the pan. Simmer for 1½ hours. Remove the bag of pips and squeeze the juice into the pan. Peel and core the apples and cut them into slices. Simmer the apples in 4 tbsp water until soft and turned to pulp. Stir the apples and oranges together and stir in the sugar over low heat until dissolved. Add the crystallized ginger cut in small pieces and the ground ginger. Boil hard to setting point. Cool for 5 minutes, stir well, pour into hot jars and cover.

• LEMON • MARMALADE

8 large lemons

4 litres/8 pints water

sugar

Peel the lemons very thinly and cut the peel into very fine shreds. Remove the white pith and pips and put into a piece of muslin and suspend in the pan. Cut up the flesh of the lemons. Put into the pan with the peel and water and simmer for 1½ hours. Remove the bag of pips and squeeze the liquid into the pan. Weigh the contents of the pan and add an equal weight of sugar. Stir in the sugar over low heat until dissolved. Bring to the boil and boil hard to setting point. Cool for 5 minutes and stir well. Pour into hot jars and cover.

• LEMON SHRED • MARMALADE

5 large lemons

1 grapefruit

3 litres/6 pints water

sugar

Peel the lemons thinly and cut the peel in fine shreds. Cover the peel with 600ml/1 pint water and simmer with the lid on until the peel is soft. Take all the white pith off the lemons and cut up the flesh. Peel the grapefruit and cut the flesh into pieces. Put the fruit in a pan with the remaining water, cover and boil gently for 1½ hours until the fruit is soft. Drain the lemon shreds and add the liquid to the fruit pulp. Bring to the boil and then strain through a jelly bag. Measure the liquid and allow 500g/1lb sugar to each 600ml/1 pint liquid. Put the sugar, liquid and lemon peel into a pan and stir over low heat until the sugar has dissolved. Boil hard to setting point. Cool for 5 minutes and stir well. Pour into hot jars and cover.

Green Tomato Chutney (page 127).

◆ ORANGE SHRED ◆
MARMALADE

1kg/2lb Seville oranges

2.5 litres/4½ pints water

2 lemons, juice

1.5kg/3lb sugar

Peel enough thin rind from the oranges to weigh 125g/4oz, and cut into thin strips. Cut up the fruit and remaining peel and put into a pan with half the water and the lemon juice. Cover and simmer for 2 hours. Simmer the thin peel in 600ml/1 pint water until soft. Strain the liquid into the fruit pulp. Strain the pulp through a jelly bag and leave to drip for 15 minutes. Add the remaining water to the fruit pulp and simmer for 20 minutes then strain through a jelly bag overnight. Mix the two liquids together and stir in the sugar over low heat until dissolved. Add the peel and boil hard to setting point. Cool for 5 minutes and stir well. Pour into hot jars and cover.

◆ PRESSURE COOKER ◆
MARMALADE

1kg/2lb Seville oranges

2 lemons, juice

1 litre/2 pints water

2kg/4lb sugar

Peel the oranges thinly and cut peel into thin strips. Take the pith from the fruit and put it into a piece of muslin with the pips. Cut up the fruit roughly. Put fruit, peel, lemon juice and the muslin bag of pips into the pressure pan with half the water. Bringing to 5kg/10lb (medium) pressure and cook for 10 minutes. Reduce pressure at room temperature. Take out the bag of pips and squeeze the liquid into the pan. Add the remaining water and sugar to the open pan and heat gently, stirring well until the sugar has dissolved. Boil hard to setting point. Cool for 5 minutes, stir well, pour into hot jars and cover.

◆ SEVILLE ORANGE ◆
MARMALADE

1kg/2lb Seville oranges

2 litres/4 pints water

1 lemon

2kg/4lb sugar

Wipe the fruit and cut the oranges in half. Squeeze out the juice and the pips. Tie the pips into a piece of muslin to suspend in the pan. Put the orange juice in the pan with the water and juice of the lemon. Slice the peel thinly and add to the pan. Simmer for about 1½ hours until the peel is soft and the liquid is reduced by half. Take out the bag of pips and squeeze out any liquid into the pan. Stir in the sugar over low heat until dissolved. Boil rapidly to setting point. Cool for about 15 minutes in the pan, stir well, pour into hot jars and cover.

DARK ◆ SEVILLE ORANGE ◆
MARMALADE

1.5kg/3lb Seville oranges

2.5 litres/5 pints water

1 lemon

3kg/6lb sugar

1 tbsp black treacle

Wipe the fruit and cut the oranges in half. Squeeze out the juice and pips. Tie the pips into a piece of muslin to suspend in the pan. Put the juice in a pan with the water and juice of the lemon. Cut the fruit into thick shreds and add to the pan. Simmer for about 1½ hours until the peel is tender. Take out the bag of pips and squeeze out any liquid into the pan. Stir in the sugar and black treacle over low heat until dissolved, and then boil quickly to setting point. Cool for about 15 minutes in the pan, stir well, pour into jot jars and cover. Four tbsp rum or whisky may be stirred into the marmalade just before pouring into pots.

Rhubarb Chutney (page 133).

• TANGERINE • MARMALADE

1kg/2lb tangerines
1 grapefruit
1 lemon
2.5 litres/5 pints water
2 tsp tartaric acid
1.5kg/3lb sugar

Peel the tangerines and cut the peel into fine shreds. Put into a pan with 600ml/1 pint water and simmer for 30 minutes. Peel the grapefruit and lemon and mince the peel. Cut up the tangerines, grapefruit and lemon flesh and put into a pan with the minced peel, water and acid. Simmer for 1½ hours. Strain the tangerine peel and add the liquid to the fruit pulp. Bring to the boil and strain through a jelly bag. Return the juice to the pan and stir in the sugar over low heat until it has dissolved. Add the tangerine peel and boil quickly to setting point. Cool for 5 minutes, stir well, pour into hot jars and cover.

• THREE FRUIT • MARMALADE

2 grapefruit
2 sweet oranges
4 lemons
3 litres/6 pints water
3kg/6lb sugar

Cut all the fruit in half. Take out the pips. Remove the pith and membranes from the grapefruit. Put the pips, pith and membranes into a piece of muslin and suspend in the pan. Shred all the fruit peel finely and cut up the flesh roughly. Put the peel, flesh and water into a pan with the bag of pips and simmer for 1½ hours. Take out the bag of pips and squeeze the juice into the pan. Stir in the sugar over low heat until dissolved and boil rapidly to setting point. Cool for 5 minutes, stir well, pour into hot jars and cover.

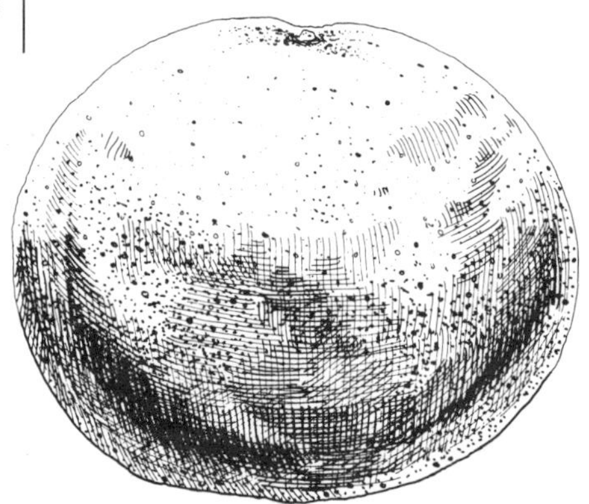

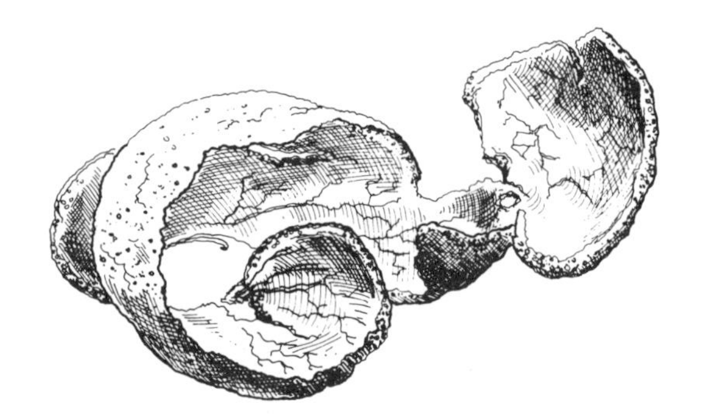

◆ JELLY ◆

Jelly-making is a little more complicated than jam-making, but the results are very rewarding. Careful measurements and attention to detail are important in jelly-making, and the only special equipment necessary is a jelly bag. A specially-made flannel bag can be obtained at good hardware shops, but a bag can be made from a double layer of muslin or from a clean, boiled, tea-towel tied with string. The jelly bag can be suspended on a special stand or between the legs of an upturned chair.

The perfect jelly should be sparkling clear with a bright colour and fresh fruit flavour. As well as being useful for spreads for bread, toast or biscuits, jellies make good tart-fillings, and can be used to glaze fruit tarts. Many jellies are excellent served with meat, poultry or game.

Wash all fruit, which should be ripe, but not over-ripe. Hard fruit such as apples and quinces should be sliced without peeling and coring; stone fruit should be cut in half, and soft fruit can be left on stems if these are clean. Crush the fruit lightly in the pan with a spoon to start the juices flowing. A small proportion (about one-third) of slightly under-ripe fruit will help to give a good set, and apples, redcurrants and gooseberries are rich in pectin, so may be mixed with other fruit such as raspberries and strawberries which do not set well on their own.

Some berries and currants need no water, but blackcurrants and hard fruits need water to help soften their skins; generally hard fruit should be covered with water. If fruit has no additional water, cook it very gently so that juices run but do not dry out. When the fruit is soft and the juices extracted, the pulp should go into the jelly bag to drip slowly into a bowl, and the process may take an hour or two, or overnight. Never be tempted to squeeze, stir or shake the fruit pulp, or the jelly will be cloudy.

The liquid must be measured, and usually 500g/1lb sugar is allowed to each 600ml/1 pint juice for fruit rich in pectin and acid, but 350g/12oz sugar will be enough for fruit with a poorer setting quality. If the sugar is warmed first, it will speed up the dissolving process, but otherwise, the sugar can be slowly heated and dissolved in the liquid. Apple and gooseberry jellies keep a better colour if cold sugar is added to cold juice.

Once the sugar has been dissolved, the jelly must be boiled rapidly to setting point, which is reached at 104°C/220°F, when the jelly will partly set on the spoon, and drops run together to form flakes which drop cleanly from the spoon. Skim jelly with a metal spoon dipped in boiling water and pour into small hot jars, tilting the jars so that the jelly does not form air bubbles. Cover with waxed circles at once, and cover completely at once or when cold, but do not move the jelly until it has set. Be sure to store it in a cool dry place.

• APPLE JELLY •

3kg/6lb cooking apples

2 lemons, juice

sugar

Wash the apples and remove any bruised or damaged pieces before weighing. Do not peel or core the apples, but cut them into pieces and put into a pan with the juice of the lemons and enough water to cover. Simmer until the apples are soft and the liquid is reduced by about one-third. Strain through a jelly bag. Measure the liquid and allow 500g/1lb sugar to 600ml/1 pint liquid. Add the cold sugar to the cold liquid and bring to the boil, stirring until the sugar has dissolved. Boil hard to setting point. Skim, pour into small hot jars and cover. For a rich red colour, add a few blackberries, cranberries, redcurrants or raspberries to the apples. If two rose-geranium leaves are simmered in the apple pulp, the apple jelly will have a deliciously scented flavour.

Use apple jelly as the basis of herb jelly to eat with meat (*see* Herb Jelly, page 102). A mixture of apples can be used for jelly, and windfalls can be used up, but damaged and bruised pieces must be carefully cut away.

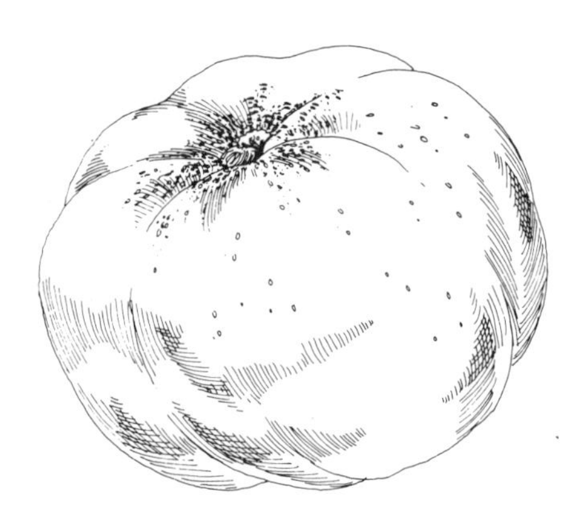

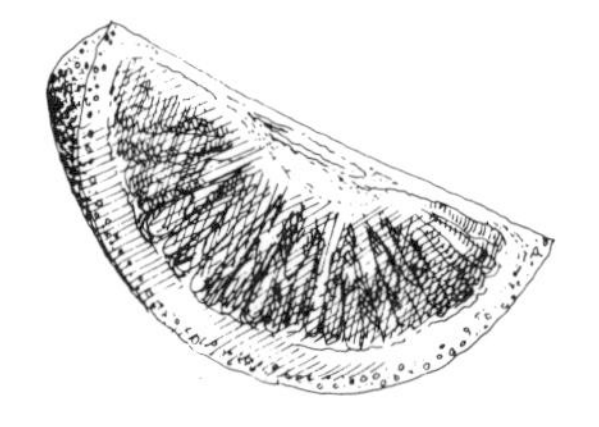

APPLE & •BLACKBERRY JELLY•

2kg/4lb blackberries

1kg/2lb cooking apples

1 litre/2 pints water

sugar

Wash the blackberries and remove any stems and unripe berries. Put the blackberries into a pan. Wash the apples and cut them up without peeling or coring. Mix with the blackberries and water and simmer for about an hour until the fruit is soft. Strain through a jelly bag and measure the juice. Allow 500g/1lb sugar to each 600ml/1 pint juice. Heat the juice gently, stirring in the sugar until dissolved. Boil hard to setting point. Pour into small hot jars and cover.

• AUTUMN JELLY •

1.5kg/3lb elderberries

500g/1lb cooking apples

500g/1lb damsons

500g/1lb blackberries

1 litre/2 pints water

1 tsp ground cloves

1 tsp ground allspice

½ tsp ground ginger

pinch cinnamon

sugar

Pick the elderberries from their stalks. Do not peel or core the apples, but cut them in pieces. Mix with the elderberries, damsons and blackberries in a pan with water and spices. Simmer for about 1 hour until the fruit is soft. Strain through a jelly bag and measure the juice. Allow 500g/1lb sugar to each 600ml/1 pint juice. Heat the juice gently, stirring in the sugar until dissolved. Boil hard to setting point. Pour into small hot jars and cover.

• BLACKCURRANT • JELLY

2kg/4lb blackcurrants

1.5 litres/3 pints water

sugar

Wash the blackcurrants and remove any stems and leaves. Simmer the blackcurrants in water for 1 hour until soft. Strain through a jelly bag and measure the juice. Allow 500g/1lb sugar to each 600ml/1 pint juice. Heat the juice gently, stirring in the sugar until dissolved. Boil hard to setting point. Pour into small hot jars and cover.

• BRAMBLE • (BLACKBERRY) JELLY

2kg/4lb blackberries

2 lemons, juice

300ml/½ pint water

sugar

Use slightly under-ripe berries for this jelly. Wash them well and discard any unripe fruit and stems. Put into a pan with the lemon juice and water and simmer for 1 hour until the fruit is soft. Strain through a jelly bag and measure the juice. Allow 500g/1lb sugar to each 600ml/pint juice. Heat the juice gently, stirring in the sugar until dissolved. Boil hard to setting point. Pour into small hot jars and cover. If liked, a pinch each of ground mace, nutmeg and cinnamon may be added to the jelly, and this spiced jelly is delicious served with cold meat.

• CRAB-APPLE • JELLY

2kg/4lb crab-apples

1 litre/2 pints water

6 cloves

sugar

Wash the fruit and remove any damaged or bruised pieces. Do not peel or core the fruit, but cut it in quarters. Put into a pan with the water and cloves. Bring to the boil and then simmer until the apples are very soft. It may be necessary to add a little more water if the fruit begins to boil dry. Strain through a jelly bag and measure the juice. Allow 500g/1lb sugar to each 600ml/1 pint of juice. Heat the juice gently and stir in the sugar until dissolved. Boil hard to setting point. Pour into small hot jars and cover.

• CRANBERRY • JELLY

1kg/2lb cranberries

600ml/1 pint water

sugar

Cook the cranberries in the water very gently until the fruit is tender. Strain through a jelly bag and measure the juice. Allow 500g/1lb sugar to each 600ml/1 pint juice. Heat the juice gently, stirring in the sugar until dissolved. Boil hard to setting point. Pour into small hot jars and cover. This is particularly good served with turkey, chicken or game.

• GOOSEBERRY • JELLY

2kg/4lb green gooseberries

sugar

Wash the berries and top and tail them. Put the fruit into a pan with just enough water to cover. Simmer for about an hour until the fruit is soft. Strain through a jelly bag and measure the juice. Allow 500g/1lb sugar to each 600ml/1 pint of juice. Add the cold sugar to the cold juice, and bring to the boil, stirring until the sugar has dissolved. Boil hard to setting point. Pour into small hot jars and cover. A flavour of muscat grapes will be imparted to the jelly if a bunch of elderflower heads is tied into muslin and hung in the juice while it is being cooked with the sugar. Gooseberry jelly, like apple jelly, is a perfect basis for herb jellies (opposite), and has a pretty green colour. It may be easier to make herb jellies when the gooseberries are in season and herbs are abundant and fresh, rather than waiting for later apples.

• GREEN GRAPE • JELLY

unripe grape thinnings

sugar

This is a useful and delicious way of using thinnings from a vine. The grapes should be about the size of peas. Wash them and cover with water. Simmer until soft and strain through a jelly bag. Measure the juice and allow 500g/1lb sugar to each 600ml/1 pint juice. Heat the juice gently, stirring in the sugar until dissolved. Boil hard to setting point. Pour into small hot jars and cover.

• HERB JELLY •

cooking apples

white vinegar

fresh mint

sugar

green food colouring

Wash the apples but do not peel or core them. Cut them in pieces and put into a pan. Just cover the apples with liquid, using 600ml/1 pint water to 150ml/¼ pint white vinegar. Add a large bunch of washed mint and simmer until the fruit is very soft. Strain through a jelly bag and measure the juice. Allow 500g/1lb sugar to each 600ml/1 pint liquid. Stir in the sugar over low heat until dissolved, and then boil hard to setting point. Just before cooking is completed, add some finely chopped fresh mint and a few drops of green food colouring. Skim well, cool slightly and stir well. Pour into small hot jars and cover. Serve with roast lamb.

Variations

1. *Alternative Herb Jelly*
Prepare jelly using 2kg/4lb gooseberries or redcurrants with water and herbs, but omitting vinegar. Finish as for herb jelly made with apples.
2. *Sage, Parsley, Thyme and Bay Leaf Jelly*
Make as for Mint Jelly, with either apples or gooseberries. Sage, parsley and thyme jellies are excellent with duck, pork or ham. Bay leaf jelly is good served with fish and chicken, but should be made without vinegar, and no bay leaves should be added when boiling liquid and sugar.

• JAPONICA JELLY •

1.5kg/3lb japonica fruit

4 tbsp lemon juice

3 litres/6 pints water

sugar

Japonica jelly is similar to quince jelly in flavour. Wash the fruit but do not peel or core. Cut them into pieces and put into a pan with the lemon juice and water. Simmer for about an hour until the fruit is soft. Strain through a jelly bag and measure the juice. Allow 500g/1lb sugar to each 600ml/1 pint juice. Heat the juice gently, stirring in the sugar until dissolved. Boil hard to setting point. Pour into small hot jars and cover. This tastes particularly delicious with milk puddings.

• LEMON JELLY •

6 large lemons

1.5 litres/3 pints water

sugar

Wipe the lemons but do not peel them. Slice them thinly and remove the pips. Put the fruit into a pan with the water. Tie the pips into a piece of muslin and suspend the bag in the pan. Bring to the boil and simmer for 1½ hours. Strain through a jelly bag and measure the liquid. Allow 500g/1lb sugar to each 600ml/1 pint liquid. Heat the lemon liquid to boiling point, and then stir in the sugar until dissolved. Boil hard to setting point. Skim well, pour into small hot jars and cover.

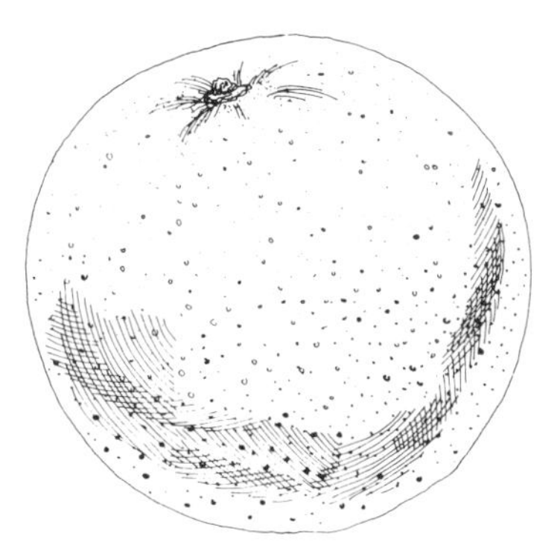

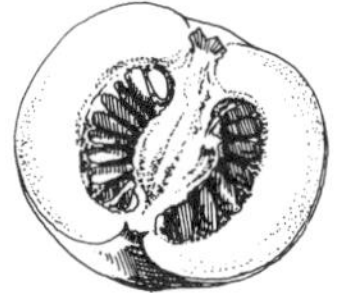

• MEDLAR JELLY •

1kg/2lb medlars

1 lemon

sugar

Use very, soft ripe medlars. Peel them and remove the pips. Slice the flesh into a pan with enough water to cover the fruit. Cut up the lemon without peeling, and add to the pan. Simmer until the fruit is soft. Strain through a jelly bag and measure the juice. Allow 350g/12oz sugar to each 600ml/1 pint juice. Heat the juice gently, stirring in the sugar until dissolved. Boil hard to setting point. Skim well, pour into small hot jars and cover.

• ORANGE JELLY •

500ml/1 pint orange juice

2 lemons, juice

250ml/½ pint water

875g/1¾lb sugar

Strain the orange and lemon juice. Put into a pan with the water and boil for 10 minutes. Stir in the sugar over low heat until dissolved. Boil hard to setting point. Pour into small hot jars and cover.

• QUINCE JELLY •

2kg/4lb quinces

3 litres/6 pints water

sugar

Wash the quinces but do not peel or core them. Cut them in small pieces and put into a pan with two-thirds of the water. Cover and simmer for 1 hour until the fruit is soft. Strain the liquid and keep it. Add the remaining water to the pulp and simmer again for 30 minutes. Strain the liquid and mix the two liquids together. Allow 500g/1lb sugar to each 600ml/1 pint liquid. Bring the juice to the boil, stir in the sugar and bring back to the boil. Boil rapidly to setting point. Skim well, pour into small hot jars and cover.

RASPBERRY & REDCURRANT JELLY

1kg/2lb redcurrants

1kg/2lb raspberries

600ml/1 pint water

sugar

Strip the redcurrants from their stems. Put the fruit into the water and simmer gently until the fruit is very soft. Strain through a jelly bag and measure the juice. Allow 500g/1lb sugar to each 600ml/1 pint juice. Heat the juice gently, stirring in the sugar until dissolved. Boil hard to setting point. Pour into small hot jars and cover. The redcurrants ensure that the jelly will have a good set but the flavour of raspberries is retained. This is a good jelly for glazing fruit flans.

REDCURRANT JELLY

1.5kg/3lb redcurrants

600ml/1 pint water

sugar

Strip the redcurrants from their stems. Put the fruit and water in a pan and simmer gently until the fruit is very soft. Strain through a jelly bag and measure the juice. Allow 500g/1lb sugar to each 600ml/1 pint juice. Stir in the sugar over low heat until dissolved and then boil hard to setting point. Pour into small hot jars and cover, but work quickly as the jelly sets very rapidly. Serve with lamb, hare or game.

Variation

Spiced Redcurrant Jelly. Make in the same way as redcurrant jelly, but add 150ml/$\frac{1}{4}$ pint white vinegar to the liquid. Tie $\frac{1}{2}$ cinnamon stick and 3 cloves into a piece of muslin to cook with the fruit, and remove when the fruit is put into the jelly bag. Serve with lamb or game.

ROWANBERRY JELLY

2kg/4lb rowanberries

4 tbsp lemon juice

750ml/1$\frac{1}{4}$ pints water

sugar

Remove ripe rowanberries from their stems, and put them into a pan with the lemon juice and water. Simmer for about 45 minutes until the fruit is soft. Strain through a jelly bag and measure the juice. Allow 500g/1lb sugar to each 600ml/1 pint juice. Heat the juice gently, stirring in the sugar until dissolved. Boil hard to setting point. Pour into small hot jars and cover. Serve with lamb, game and venison.

STRAWBERRY & REDCURRANT JELLY

500g/1lb strawberries

250g/8oz redcurrants

4 tbsp water

sugar

Put the strawberries and redcurrants into a pan with the water and simmer gently until the fruit is soft. Strain through a jelly bag and measure the juice. Allow 500g/1lb sugar to each 600ml/1 pint juice. Heat the juice gently, stirring in the sugar until dissolved. Boil hard to setting point. Pour into small hot jars and cover.

• SUMMER JELLY •

500g/1lb redcurrants

500g/1lb raspberries

500g/1lb strawberries

500g/1lb black cherries

1 tsp tartaric acid

600ml/1 pint water

sugar

Use fruit which is firm and just ripe. Put into a pan with the tartaric acid and water and simmer until the fruit is soft. Strain through a jelly bag and measure the juice. Allow 500g/1lb sugar to each 600ml/1 pint juice. Heat the juice gently and stir in the sugar until dissolved. Boil hard to setting point. Pour into small hot jars and cover. This jelly is particularly good for tarts.

• TOMATO JELLY •

1.5kg/3lb red tomatoes

3 tbsp lemon juice

½ cinnamon stick

750g/1½lb/ sugar

Wipe the tomatoes and cut them in pieces. Simmer very gently until soft. Strain through a jelly bag. Heat the juice with the lemon juice and cinnamon to boiling point. Stir in the sugar until dissolved. Boil hard to setting point and take out the cinnamon. Pour into small hot jars and cover. Serve with cream cheese, ham or poultry.

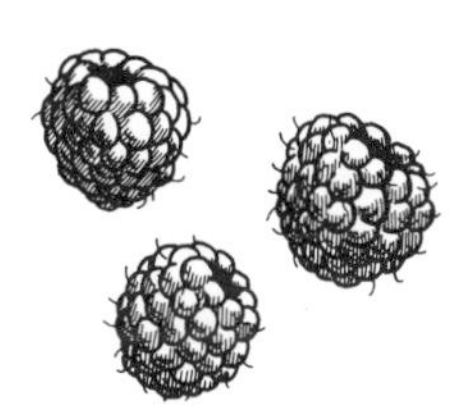

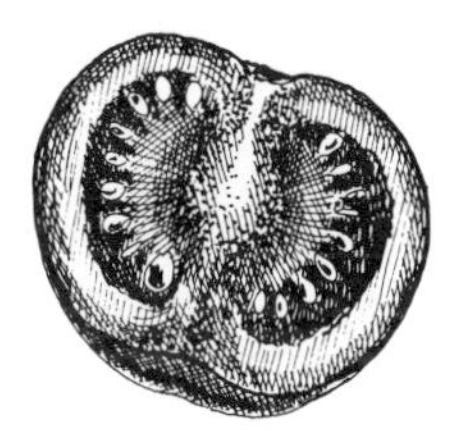

• FRUIT CURD •

Curds are creamy fruit mixtures made with fresh fruit, eggs, butter and sugar which have a short storage life, but which can be frozen to extend this life. They are best made in small quantities and packaged in small jars (for cupboard storage – 2 months) or small freezer containers (6 months). Curds should be made with cube or caster sugar and unsalted butter, together with fresh fruit and eggs. The beaten eggs are best strained before adding to the mixture to give a smooth texture. Curds should be cooked in a double saucepan or in a bowl over hot water, and the cooking heat should be low. The mixture must be stirred well during cooking and will be creamy and coat the back of a spoon when ready. Curds thicken as they cool. They are delicious for tart and cake fillings and spreads, or may be used as sauces for ices or puddings.

Curds must be covered with a waxed disc and cellophane cover for ordinary storage but do not use a twist top. For freezer preservation, they may be packed in freezer containers.

• FRESH APRICOT • CURD

250g/8oz fresh apricots

1 lemon

50g/2oz butter

250g/8oz caster sugar

2 eggs

Wash the fruit and put into a pan with just enough water to prevent burning. Cook until soft. Sieve and put into a bowl, or the top of a double saucepan. Add the grated rind and juice of the lemon, butter, sugar, and well-beaten eggs. Cook gently over hot water stirring well until the mixture thickens, which will take about 30 minutes. Pour into small hot jars and cover, or else cool and pour into small freezer containers.

BLACKBERRY & • APPLE CURD •

500g/1lb cooking apples

1kg/2lb blackberries

2 lemons, juice

250g/8oz butter

1.25kg/2½lb caster sugar

4 eggs

Peel and core the apples. Wash the blackberries and discard any stems and unripe berries. Cook the apples and berries in very little water until soft, and then sieve. Put into a double saucepan or a bowl over hot water and add the lemon juice, butter and sugar. When the butter and sugar have melted, add the well-beaten eggs and cook until the mixture thickens, stirring well. Pour into small hot jars and cover, or else cool and pour into small freezer containers.

• GOLDEN CURD •

50g/2oz butter

2 oranges

1 lemon

250g/8oz sugar

4 eggs

Melt the butter in the top of a double saucepan, or a bowl over hot water. Grate the orange and lemon rinds finely. Squeeze out the juice and strain it. Add the rinds, strained juice and sugar to the butter, and stir until the sugar dissolves. Add the well-beaten eggs and continue cooking gently until the mixture thickens and coats the back of a spoon. Pour into small hot jars and cover, or else cool and pour into small freezer containers.

• GOOSEBERRY • CURD

1.5kg/3lb green gooseberries

450ml/¾ pint water

750g/1½lb caster sugar

125g/4oz butter

4 eggs

Wash the gooseberries and top and tail them. Cook in the water until soft, and sieve. Put into a double saucepan or bowl over hot water with the sugar and butter. When the sugar and butter have melted, stir in the well-beaten eggs, and continue cooking gently, stirring until the mixture thickens. Pour into small hot jars and cover, or else cool and pour into small freezer containers.

• LEMON CURD •

4 large lemons
175g/6oz butter
500g/1lb sugar
4 eggs

Grate the lemon rinds finely, avoiding all pith. Squeeze out the juice and strain it. Melt the butter in a double saucepan or bowl over hot water, and add the lemon rind, juice and sugar. When the sugar has dissolved, stir in the well-beaten eggs. Cook gently, stirring well until the mixture thickens. Pour into small hot jars and cover, or else cool and pour into small freezer containers.

• ORANGE PEEL • CURD

1 large orange
50g/2oz candied orange peel
125g/4oz sugar
125g/4oz butter
3 egg yolks

Grate the orange rind finely, avoiding all pith. Squeeze out the juice and strain it. Chop the candied peel finely. Put the sugar and butter in a double saucepan or in a bowl over hot water, and add the rind, juice and chopped peel. When the sugar and butter have melted, stir in the beaten egg yolks. Continue cooking gently and stirring well until the mixture thickens. Pour into small hot jars and cover, or else cool and pour into small freezer containers.

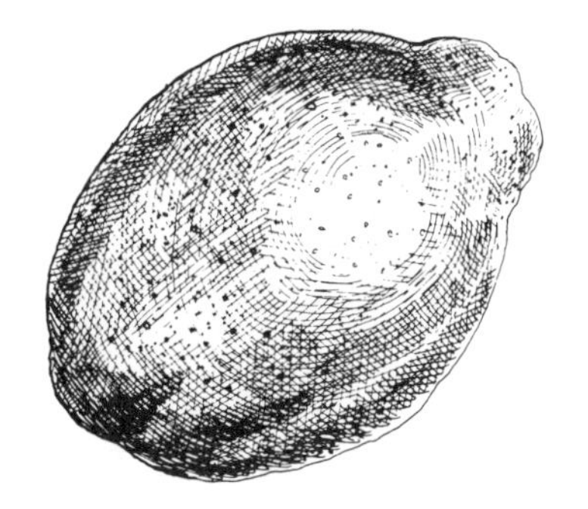

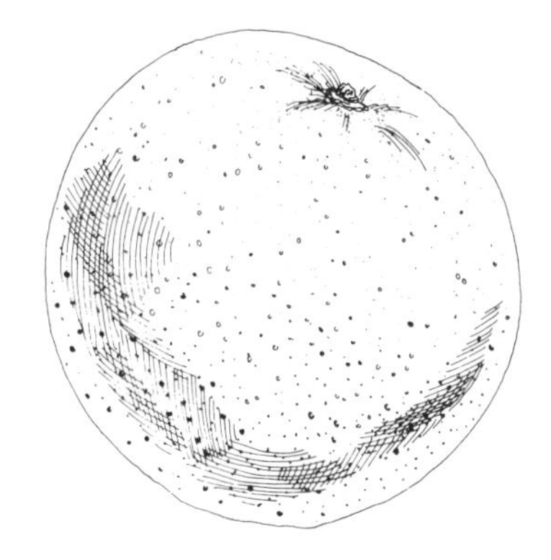

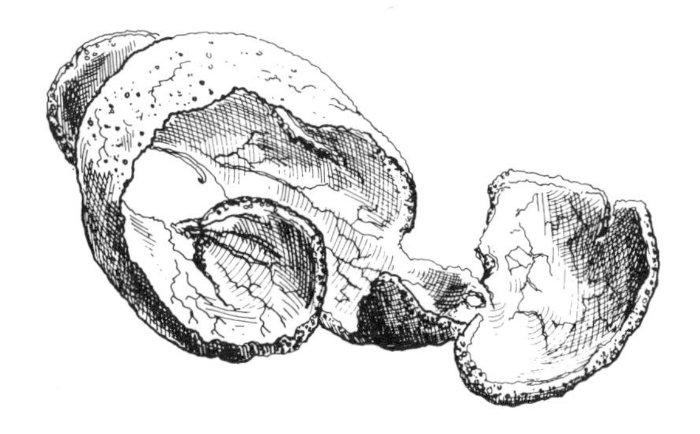

• FRUIT BUTTER & CHEESE

Fruit butter and cheese are both thick mixtures of fruit pulp and sugar, but they vary slightly. A fruit butter should be thick but not completely set, with the consistency of thick cream so that it can be spread on bread or toast. Fruit cheese is much firmer and should be put into a straight-sided jar from which it can be turned out and sliced to eat with cream or milk puddings, or with savoury dishes.

Large quantities of fruit or windfalls can be used up in making 'butter' or 'cheese', as these quantities reduce considerably during cooking. Fruit butter usually contains half the amount of sugar and fruit pulp. The pulp has to be cooked until thick before the sugar is added. Brown sugar gives a good flavour, but darkens the preserve. Ground spices also darken the preserve, so it might be better to use whole spices tied in muslin. A little lemon juice will sharpen and improve fruit flavours.

The preserves should be tested on a plate for setting point. Fruit butter is ready when no rim of liquid appears round the edge of the mixture. Fruit cheese should be much firmer and will be ready when a spoon drawn across the bottom of the pan leaves a clear line. Both butter and cheese should look thick and glossy. Put them into small jars which can be used up quickly, and use wide-mouthed jars for cheese so that it can be turned out.

• CHERRY BUTTER •

2kg/4lb cherries

1kg/2lb sugar

1 lemon

Stone the cherries, and open a few of the stones. Take out the kernels, blanch and skin. Arrange the cherries in layers in a bowl with sugar. Add the grated rind and juice of the lemon. Leave overnight. Turn the mixture into a pan, simmer for 20 minutes and then add the kernels. Boil quickly until very thick, stirring well. Pour into small hot jars and cover.

• CIDER APPLE • BUTTER

3kg/6lb apples

1 litre/2 pints water

1 litre/2 pints dry cider

granulated or soft brown sugar

1 tsp ground cloves

1 tsp ground cinnamon

1 tsp ground nutmeg

Wash the apples and cut them up without peeling. Put into a large pan with the water and cider and simmer until soft. Sieve and weigh the pulp. Simmer the pulp until thick and creamy. Add 350g/12oz sugar to each 500g/1lb of weighed pulp. Stir the sugar and spices into the apples and cook gently, stirring often, until no surplus liquid remains. Pour into hot jars and cover. A mixture of apples may be used, and the same recipe may be used for crab-apples.

• BLACKBERRY • CHEESE

2kg/4lb blackberries

2 tsp citric or tartaric acid

sugar

Wash the blackberries and remove any stems and unripe fruit. Put into a pan with just enough water to cover, and the acid. Bring to the boil and simmer gently until the fruit is soft. Put though a sieve and weigh the pulp. Allow 500g/1lb sugar to each 500g/1lb fruit pulp and stir in the sugar over low heat until dissolved. Bring to the boil and then cook gently until thick, stirring well. Pour into small hot, wide-necked jars, and cover.

• GOOSEBERRY • CHEESE

1.5kg/3lb green gooseberries

300ml/½ pint water

sugar

Top and tail the gooseberries and wash them. Simmer in the water until soft. Put through a sieve and weigh the pulp. Allow 350g/12oz sugar to each 500g/1lb fruit pulp, and stir in the sugar over low heat until dissolved. Bring to the boil and then cook gently until thick, stirring well. Pour into small hot, wide-necked jars, and cover.

• PLUM & APPLE • BUTTER

1.5kg/3lb apples

500g/1lb plums

granulated sugar

Peel and core the apples and cut them up. Cook in very little water until soft. Wash the plums and cut them up, taking out the stones. Add the plums to the apples and cook until soft. Sieve and weigh the fruit, and add 350g/12oz sugar to each 500g/1lb fruit pulp. Stir the sugar into the fruit over low heat until the sugar has dissolved. Boil until thick and creamy. Pour into hot jars and cover.

• RHUBARB BUTTER •

1kg/2lb rhubarb

150ml/¼ pint water

500g/1lb sugar

red food colouring

Wash the rhubarb and cut it in pieces. Put into a pan with the water and simmer until tender. Sieve and return to the pan. Bring to the boil, stir in the sugar, and cook over low heat, stirring well until the sugar has dissolved. Cook and stir until the mixture is thick and creamy. Add a little red food colouring. Pour into hot jars and cover.

• SLOE CHEESE •

2kg/4lb sloes

sugar

Sloes are small, dark wild plums which make a delicious, and slightly sharp, cheese. Wash the fruit and then simmer in very little water until soft. Sieve and weigh the pulp and allow 500g/1lb sugar to each 500g/1lb of fruit pulp. Stir in sugar over low heat until dissolved and then bring to the boil. Cook gently, stirring well until thick. Pour into small, hot, wide-necked jars, and cover.

♦ MINCEMEAT ♦

Fruit mincemeat is a favourite preserve with a long history. It was originally made with minced meat (sometimes tongue) mixed with dried fruit and a lot of spice, together with apples and sugar, and a preservative spirit. Many of the old ingredients remain, but the meat has been replaced by shredded beef suet. The mincemeat looks best if it is composed of a mixture of minced, chopped and whole fruit, and it should be packed into clean, dry, cold jars. A plastic or twist top will prevent evaporation and drying-out, but if the mixture does become a little dry, extra spirits may be stirred in before use. The flavours mature after a week or two, so be sure to make mincemeat well ahead of the time it will be needed.

♦ BASIC FRUIT ♦ MINCEMEAT

250g/8oz cooking apples
250g/8oz shredded beef suet
250g/8oz raisins
250g/8oz currants
250g/8oz sultanas
125g/4oz chopped mixed peel
1 lemon
125g/4oz soft brown sugar
2 tbsp brandy or sherry

Peel and core the apples. Put the apples, suet, dried fruit and peel through the coarse screen of a mincer. Mix in the grated rind and juice of the lemon, sugar and brandy or sherry. Stir very thoroughly, pack into clean cold jars and cover at once.

♦ CIDER ♦ MINCEMEAT

350g/12oz cooking apples
350g/12oz stoned raisins
175g/6oz currants
175g/6oz soft brown sugar
2 tsp ground cinnamon
1 tsp ground cloves
1 tsp ground nutmeg
75ml/3fl. oz dry cider
1 lemon
75g/3oz butter
1 tbsp brandy

Peel and core the apples. Chop them in small pieces and put them into a pan. Use large raisins and cut out the stones. Put into the pan with the apples, currants, sugar, spices, cider, grated rind and lemon juice, and butter. Simmer for 30 minutes, stirring well. Remove from the heat and stir in the brandy. Pack into jars and cover when cold. This is a short-keeping mincemeat which is best used within two weeks, but it can be kept longer in the refrigerator if necessary.

◆ RICH FRUIT ◆ MINCEMEAT

1kg/2lb cooking apples

350g/12oz shredded beef suet

250g/8oz raisins

250g/8oz sultanas

500g/1lb currants

250g/8oz chopped mixed peel

1 lemon

15g/½oz ground cinnamon

750g/1½lb soft brown sugar

2 tsp salt

150ml/¼ pint brandy

150ml/¼ pint sherry

Peel and core the apples. Put them through the coarse screen of the mincer with the suet, dried fruit and peel. Sir in the grated rind and juice of the lemon with the other ingredients. Leave in a cold place for 12 hours. Stir well and leave for another 12 hours. Pack into clean cold jars and cover at once. This mincemeat will improve if kept for 6 weeks before use.

◆ RUM ◆ MINCEMEAT

500g/1lb cooking apples

750g/1½lb shredded beef suet

500g/1lb currants

500g/1lb stoned raisins

500g/1lb sultanas

500g/1lb dark soft brown sugar

25g/1oz ground mixed spice

2 lemons

3 oranges

150ml/¼ pint rum

150ml/¼ pint brandy

Peel and core the apples. Put through the coarse screen of the mincer with the suet. Use large raisins and cut out the stones. Mix the currants, raisins, sultanas, sugar and spice with the apples and suet. Add the grated rind and juice of the lemons and oranges, together with the rum and brandy. Mix very thoroughly and pack into clean jars. Cover at once.

PICKLES, CHUTNEYS & SAUCES

Surplus fruit and vegetables can be turned into delicious and useful pickles, chutneys and sauces by preserving them in vinegar and sugar with spices. One particular advantage of these spicy relishes is that they can usually be made with a selection of mixed raw materials, so that the odd cauliflower, a few onions, some windfall apples or a spare vegetable marrow need never be wasted, and the resulting preserves will do a lot to liven up winter meals. As with other kinds of preserve, make a number of varieties in small quantities, rather than an enormous batch of one kind.

• EQUIPMENT •

The equipment needed for preserving in vinegar is available in most houses, but it is important that the correct items are used as acid can affect both the equipment and the finished produce. A large preserving pan or saucepan is necessary, as quantities can be large and there must be room for cooking without boiling over. Stainless steel, aluminium or unchipped enamel are suitable, but iron, brass or copper must never be used.

Blackberry Cheese (page 109).

Accurate scales should be used, as though pickle-making does not have to be quite as accurate as cake-making, results will be better if ingredients are carefully measured. A large bowl of earthenware or oven-proof glass is useful for preparing some pickles which have to be left in salt before cooking. A long-handled wooden spoon is safest for stirring, and a sharp stainless knife is best for speedy cutting.

Some recipes specify the sieving of ingredients, and this should be done through a non-metal sieve. This can also be used for straining spiced vinegar, or the vinegar can be allowed to drip through a jelly bag, but a metal strainer should never be used. Small pieces of muslin or well-boiled linen are useful for tying up whole spices to suspend in the pan.

The correct jars must be used for preserves which contain vinegar. Pickles and chutneys are best stored in wide-mouthed jars, while sauces can obviously go into bottles. Jam jars with clip-on or screw-top lids can be used for vinegar-based preserves, or old pickle jars which have been sterilized, or preserving jars. Uncoated metal must not touch the finished product, and lids should either be coated or lined with a vinegar-proof card disc. Coffee jars with plastic lids are suitable for storage, but paper or transparent jam covers are not suitable as they will allow evaporation.

Labels are essential as these preserves look very similar after storage, and pickles need to be identified and dated.

• INGREDIENTS •

Apart from fruit and/or vegetables, the main ingredients of pickles, chutneys and sauces are vinegar, sugar, spices and sometimes salt and dried fruit.

Vinegar should be of good quality and containing at least 5 per cent acetic acid (branded vinegars contain 5–7 per cent), but draught vinegar usually has a low acetic acid content. Malt vinegar has a strong flavour and colour, and white vinegar has a better appearance in clear pickles and spiced fruit. Wine vinegar and cider vinegar are more expensive but have a better flavour for more delicate preserves. When spiced vinegar is specified, this can be prepared beforehand.

Sugar is needed for many pickles, and granulated sugar will not add colouring. Demerara and soft brown sugar give both colour and flavour to chutneys and sauces.

Dried fruits give richness, colour and sweetness to chutney. Raisins and sultanas are most commonly used, but dates and apricots are both delicious in chutney.

Preparing and mincing Pickled Horseradish (page 119).

Spices are very important in pickling and give great individuality to recipes. Ground spices are suitable for chutneys and sauces as they blend in easily and flavouring can be adjusted to suit family tastes. They will however make vinegar cloudy and should not be used for preparing spiced vinegar, or for making clear pickles or spiced fruits. Whole spices should be tied in a piece of muslin which can be suspended in the pan, and it is best to crush the spices lightly with a weight first to release their flavour. Whether spices are whole or ground, they should always be fresh, as they can taste stale after long storage.

• MAKING PICKLES, CHUTNEYS • & SAUCES

Whatever the end product, the cooking procedure for vinegar-based preserves is similar. First, read a recipe carefully to see if there has to be any pre-cooking preparation which may take as long as 24 hours if brining or layering with sugar is required, and which may therefore throw out a cooking schedule. Gather all ingredients and equipment together, and weigh items carefully. Cut up fruit and vegetables with a sharp, stainless knife, after removing any bruised or damaged parts. Check that fruit is in the condition required for a recipe (i.e. ripe or slightly under-ripe) as this can affect the quality of the finished product. Follow the recipe for cooking times, and make sure that chutney is thick and richly-coloured. Always make sure jars are clean, sterilized with hot water, dried and warmed before filling. Fill carefully to avoid splashing hot ingredients on yourself and the jars, cover carefully and label. Store in a cool, dry, dark place.

Spiced Vinegar. Spiced vinegar is needed for many recipes and this may be prepared at the same time as the recipe, or up to a month beforehand so that the vinegar is more fully-flavoured with the spices. Allow 15g/½oz cinnamon stick, 15g/½oz blade mace, 15g/½oz allspice berries, 8g/¼oz peppercorns, 8g/¼oz whole cloves to each 2 litres/4 pints vinegar. Tie the spices in a piece of muslin and suspend in a pan with the vinegar. Bring slowly to simmering point with a lid on so that no flavour is lost. Remove from the heat and leave to stand for about 3 hours before use, or keep in a covered jar in a cold place for a month. Strain and use hot or cold as required. Cold vinegar keeps pickles crisp when made with uncooked vegetables.

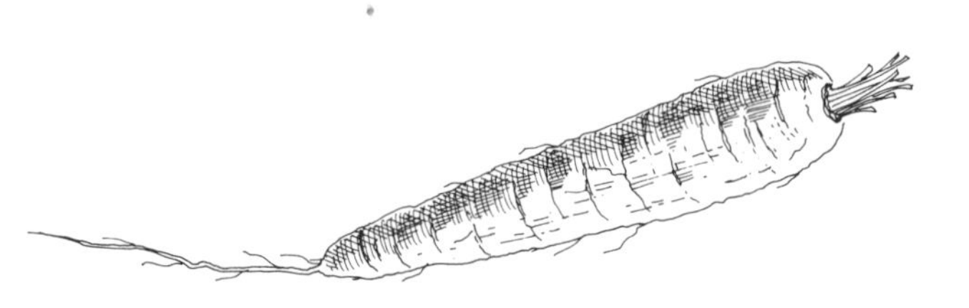

◆ Storage times

Chutney matures in storage and should be left for 2 months after making to develop a good flavour. Pickles do not need such long storage and are generally ready to eat after about 2 weeks. Some raw vegetable pickles can lose their crispness within a year, and red cabbage is only at its best for 6 months. Spiced pickled fruit is best stored for 2 months before using.

◆ Pickles

Many vegetables and fruit can be preserved in vinegar with spices. Fruit is usually sweetened as well to produce delicious and unusual pickles to eat with meat, poultry and fish. Use good vegetables and fruit of high quality and remove all blemishes, avoiding any fruit which is soft and slushy. Make up spiced vinegar well ahead of pickling if possible so that it develops a full flavour. If vegetables have to be soaked in brine before further preparation, be sure to allow plenty of time for this as it usually means the pickling process must be spread over two days.

Many vegetable pickles are made by brining raw vegetables, draining them and then packing firmly into jars. The jars are then filled with spiced vinegar and sealed tightly. When a recipe specifies using cooked vegetables, these should be packed into the jars firmly but without pressure which will spoil their shape. The vegetables should stand for an hour and any liquid should be drained off before the vinegar is poured in.

Fruit is normally lightly cooked in the sweetened spiced vinegar before draining and packing in jars. The vinegar syrup then has to be cooked until thickened and then poured over the fruit while hot.

Fruit is normally lightly cooked in the sweetened spiced vinegar before draining and packing in jars. The vinegar syrup then has to be cooked until thickened and then poured over the fruit while hot.

If pickles do not keep well, they have been poorly stored or packed in unclean jars. Under-brining with too little salt can leave too much water in the vegetables which reduces the strength of vinegar and affects its preserving qualities. If pickles ferment, grow mould or show white speckles, they should be thrown away – this may have resulted from weak vinegar, poor brining, or bad fruit or vegetables.

Brining. Raw vegetables for pickling need to be brined to draw out some of their natural liquid which will otherwise dilute the preserving vinegar. Cooking drives off liquid, so cooked vegetables need not be brined. Fruit contains acid and is not brined, but lightly cooked before pickling.

A dry brine or a wet brine can be used – vegetable marrow, red cabbage and cucumber are usually prepared in dry brine. The salt should be block salt, as ordinary table salt contains a chemical which keeps it free-running but may cloud the vinegar, and the block salt must be crushed or grated.

To prepare vegetables in dry brine, cut or shred them according to the recipe and arrange in layers with salt in a large bowl, allowing 75g/3oz salt to 750g/1½lb prepared vegetables. Salt should be used for the top layer in the bowl, and then a cover of cloth or paper should be arranged on top and the bowl put in a cold place. After 24 hours, there will be a lot of liquid in the bowl which needs draining off, and then surplus salt should be rinsed away under cold running water. The vegetables must be well drained before further processing.

To prepare vegetables in wet brine, put the prepared vegetables into a large bowl and cover with a salt solution. Allow 225g/8oz salt to each 2 litres/4 pints water. About 600ml/1 pint of brine will be enough for 500g/1lb vegetables and should cover them completely. Put a lightly weighted plate on top to keep the vegetables under the brine for the required time. Drain off the liquid completely before further processing.

CLEAR • MIXED PICKLE •

500g/1lb shallots
500g/1lb cauliflower
1 large cucumber
25g/1oz salt
1.5 litres/3 pints spiced vinegar
15g/½oz pickling spice

Peel the shallots. Divide the cauliflower into sprigs. Peel the cucumber and cut into cubes. Mix all the vegetables, put into a bowl, sprinkle with salt, and leave overnight. Drain off the liquid and pack the vegetables into preserving jars. Pour over the spiced vinegar and put a few pickling spices into each jar. Seal tightly.

• PICKLED APPLES • & PEPPERS

2kg/4lb eating apples
1kg/2lb green peppers
1.5 litres/3 pints cider vinegar
150g/5oz soft brown sugar
25g/1oz juniper berries

Peel and core the apples, and cut them into rings. Scoop out seeds from peppers and cut them into rings. Pack the apples and peppers in layers in hot preserving jars. Bring the vinegar, sugar and juniper berries to the boil and pour over the apples and peppers at once, then seal tightly. Leave for about 6 weeks before using.

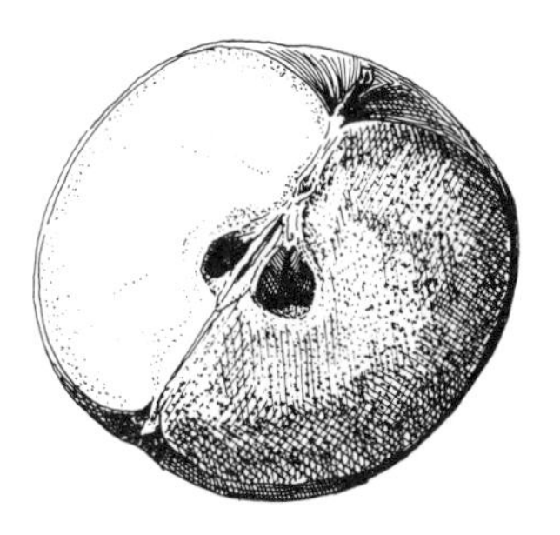

SPICED • APPLE SLICES •

1.5kg/3lb eating apples
500g/1lb sugar
600ml/1 pint white vinegar
1 tsp salt
150ml/¼ pint water
1 cinnamon stick
2 tsp cloves

Peel and core the apples, and slice them thickly. Put the sugar, vinegar, salt and water in a pan with the cinnamon and cloves tied into a piece of muslin suspended in the pan. Stir over low heat until the sugar has dissolved. Drop in the apple slices and cook gently until the fruit is tender but not broken. Lift out the apple slices with a perforated spoon and put into hot preserving jars. Bring the syrup to the boil and pour into the jars to cover the apple slices. Seal tightly.

SPICED • APRICOTS •

500g/1lb dried apricots
15g/½oz cloves
small piece of cinnamon stick
15g/½oz allspice
900ml/1½ pints white vinegar
625g/1¼lb sugar

Put the apricots in a bowl, cover with water and leave to soak overnight. Tie the spices in a piece of muslin and suspend in a pan. Add the vinegar and bring to the boil. Drain the apricots, add to the vinegar and simmer for 10 minutes. Lift out the apricots with a perforated spoon and put into hot preserving jars. Stir the sugar into the vinegar, bring to the boil and boil for 5 minutes until syrupy. Take out the bag of spices, pour the syrup over the fruit and seal tightly.

PICKLED BEETROOT

8 medium beetroot

1 litre/2 pints vinegar

15g/½oz black peppercorns

15g/½oz allspice berries

1 tsp grated horseradish

1 tsp salt

Cook the beetroot, cool, skin and slice them. Pack in preserving jars. Put the vinegar, spices, horseradish and salt into a pan, bring to the boil, and then cool. When cold, pour over the beetroot and seal at once.

Variations

1. *Beetroot and Onion Pickle*
A delicious alternative to pickled beetroot can be made by arranging layers of sliced cooked beetroot and thinly-sliced raw onions in jars and covering them with cold spiced vinegar.
2. *Beetroot and Horseradish Pickle*
After cooking, grate the beetroot and mix with two pieces grated fresh horseradish and 50g/2oz sugar. Pack into jars and cover with plain vinegar.

SPICED CRANBERRIES

2kg/4lb cranberries

450ml/¾ pint cider vinegar

150ml/¼ pint water

1.5kg/3lb sugar

25g/1oz ground cinnamon

15g/½oz ground cloves

15g/½oz ground allspice

Put all the ingredients into a pan, bring to the boil, and simmer for 45 minutes, stirring well. Pour into hot preserving jars and seal tightly.

PICKLED CUCUMBERS

1kg/2lb ridge cucumbers

1 litre/2 pints water

175g/6oz salt

50g/2oz sugar

1 litre/2 pints spiced vinegar

bay leaves

Use young cucumbers with soft skins. Wipe them, cut in half without peeling, and then in quarters lengthwise. Boil the water and salt together and leave until cold. Pour over the cucumbers and leave for 24 hours. Put the sugar and spiced vinegar into a saucepan and heat gently until the sugar has dissolved. Lift the cucumbers out of the salt brine, and rinse them in cold water. Drain and leave to dry for 2 hours. Pack upright in preserving jars, and cover with cold vinegar syrup. Put a bay leaf in each jar and seal tightly. Keep 2–3 weeks before using.

PICKLED EGGS

12 hard-boiled eggs

1 litre/2 pints white vinegar

15g/½oz root ginger

15g/½oz mustard seeds

15g/½oz white peppercorns

2 chillies

Shell the cooled eggs. Simmer the vinegar with bruised ginger, mustard seeds and peppercorns for 5 minutes, then strain and leave to cool. Arrange the eggs upright in a wide-mouthed preserving jar, and put the chillies on top. Cover with vinegar and seal tightly. Keep for 2–3 weeks before using.

PICKLED • GHERKINS •

gherkins (pickling cucumbers)

basic brine

vinegar

spiced vinegar

Gather gherkins on a dry day and wipe them well. Put in a large bowl and cover with brine. Leave for 3 days and drain off the brine. Cover the gherkins with plain boiling vinegar. Leave for 24 hours, drain off vinegar, reboil it and pour over the gherkins. Repeat the process until the gherkins are bright green. Drain off the vinegar, pack gherkins into small jars and cover with cold spiced vinegar. Seal tightly.

PICKLED • GREEN TOMATOES •

1kg/2lb green tomatoes

250g/8oz onions

125g/4oz salt

250g/8oz sugar

600ml/1 pint white vinegar

Wipe the tomatoes, and cut across in slices without peeling. Peel and slice the onions thinly. Arrange the tomatoes and onions in a bowl in alternate layers, and sprinkle layers generously with salt. Leave for 24 hours and drain off the liquid completely. Put the sugar and vinegar into a pan and heat gently until the sugar has dissolved. Put in the tomatoes and onions and simmer until soft but not broken. Put into warm preserving jars and seal at once.

PICKLED • HORSERADISH •

fresh horseradish roots

salt

vinegar

Wash the roots well in hot water and scrape off the skin. Grate or mince the flesh and pack into small jars. Use 1 tsp salt to each 300ml/½ pint vinegar, mix well and cover the horseradish. Seal tightly.

PICKLED • MUSHROOMS •

500g/1lb button mushrooms

600ml/1 pint white vinegar

1 tsp salt

½ tsp white pepper

small piece root ginger

1 small onion, sliced

Wipe the mushrooms, but do not wash or peel them, and trim stems. Put into a pan with the other ingredients. Simmer until mushrooms are tender, then lift them out with a slotted spoon and pack into preserving jars. Remove the onions and ginger. Bring the vinegar to the boil and pour over the mushrooms. Seal tightly.

• MUSTARD PICKLE •

1 medium vegetable marrow
1 medium cauliflower
1 cucumber
500g/1lb French beans
500g/1lb button onions
25g/1oz salt
300g/10oz sugar
1 litre/2 pints vinegar
50g/2oz plain flour
50g/2oz mustard
15g/½oz turmeric
15g/½oz ground ginger
15g/½oz ground nutmeg

Chop the marrow into small chunks without peeling, but remove seeds and pith. Divide the cauliflower into small pieces, chop the cucumber and beans, and peel the onions. Mix all the vegetables together in a bowl and sprinkle with salt. Cover with water, leave to soak overnight, then drain off the water. Mix all the dry ingredients with a little of the vinegar to a smooth paste. Put the vegetables into a pan with the remaining vinegar and simmer until just tender. Add a little of the boiling vinegar to the blended dry ingredients, then return to the pan of vegetables. Simmer for 10 minutes, stirring all the time, until thick. Put into preserving jars while hot.

PICKLED • NASTURTIUM • SEEDS

fresh nasturtium seeds
250g/8oz salt
1.5 litres/3 pints water
spiced vinegar
3 sprigs tarragon

Gather nasturtium seeds while still green, on a dry day. Make a brine with salt and water and cover the nasturtium seeds for 24 hours. Drain well and pack in small jars. Put into an oven at 170°C/325°F/gas mark 3 for 10 minutes. Meanwhile, heat the spiced vinegar. Take out the jars, put a few leaves of tarragon into each one, and cover with vinegar. Seal tightly.

PICKLED • ONIONS •

1kg/2lb small onions or shallots
40g/1½oz salt
1 litre/2 pints spiced vinegar

Peel the onions or shallots and put on a shallow dish. Sprinkle with salt and leave overnight. Rinse the onions or shallots under cold water and drain well. Pack into preserving jars, arranging with the handle of a wooden spoon so that there are no large spaces. Fill the jars with cold spiced vinegar and seal tightly. Keep 3–4 weeks before using.

PICKLED
• PRUNES •

500g/1lb large prunes
450ml/¾ pint cold tea
600ml/1 pint white vinegar
500g/1lb sugar
2.5cm/1in cinnamon stick
1 tsp cloves
10 allspice berries
blade of mace

The tea should be from the pot, without milk. Put the prunes into a bowl, cover with tea and leave to soak overnight. Simmer in the tea until the prunes are plump. Boil the vinegar, sugar and spices for 5 minutes, then add the prunes and cooking liquid and simmer for 5 minutes. Lift out the prunes with a slotted spoon and pack into preserving jars. Boil the syrup, pour over the prunes, and seal at once.

PICKLED
• RED CABBAGE •

1kg/2lb red cabbage
125g/4oz salt
900ml/1½ pints spiced vinegar

Remove the coarse outer leaves from the cabbage and the coarse stem. Shred the cabbage finely, put in a large bowl in layers with the salt, cover and leave for 24 hours. Drain the liquid from the cabbage and rinse under cold running water. Drain thoroughly and pack loosely in clean cold jars, filling them half-full. Cover with cold spiced vinegar. Press down the cabbage slightly and fill the jar up with more cabbage. Cover with vinegar and seal. Keep for at least a week before using.

PICKLED
• WALNUTS •

green walnuts
2 litres/4 pints water
250g/8oz salt
2 litres/4 pints vinegar
3 small onions
1 tsp cloves
40g/1½oz black peppercorns
25g/1oz root ginger
2 blades mace
2 bay leaves
50g/2oz mustard seed

Use the walnuts before any shell has formed, about the middle of July. Pierce each nut with a darning needle (this can stain the hands badly) and put into a brine made from the water and salt. Leave in the brine for 9 days, changing the brine three times. Drain and put on flat dishes, and leave in an airy place for 2–3 days until black. Boil the vinegar, onions and spices together for 5 minutes. Leave covered for 2 hours, strain and boil again. Pack the black walnuts into preserving jars, and cover with hot spiced vinegar. Leave until cold and seal tightly. Leave for 2 months before using.

• CHUTNEY •

Chutney is a mixture of fruit and/or vegetables cut into small pieces and cooked with vinegar and sugar to make a spicy product with the consistency of jam. Fruit and vegetables may be bruised and imperfect, but never mouldy or rotten, and all bruises and poor parts should be cut away. The fruit and vegetables are best cut in small pieces, but some people prefer them minced. Dried fruit, onions, garlic and brown sugar all give added flavour and richness to chutney.

Spices are particularly important in chutney and may be varied according to family tastes. Cayenne and chillies are very hot and may be omitted from recipes, but ginger which also gives 'hotness' is always acceptable in chutney. Cloves, cinnamon, nutmeg and mace are aromatic spices and particularly good for mild fruit chutneys. Use freshly ground spices and measure them accurately for the first batch of chutney, but vary quantities if liked after testing the finished product when it is cold.

Most chutneys are made by simmering together fruit and vegetables in some vinegar with spices to soften the ingredients, break down fibres and bring out flavours before adding sugar and then cooking is continued slowly until the mixture is of a thick jam-like consistency. Chutney will thicken slightly when cool but should never be bottled while pale and liquid. Stir chutney carefully, and never leave it for long periods to cook unwatched, as it will spoil if burned.

Put chutney into hot clean jars with vinegar-proof lids, and never cover with paper as the mixture will dry out. Fill jars to the brim, and wipe them well before storing in a cool, dry, dark place. If chutney becomes dry and brown on top, it has been kept without an airtight cover, or the storage place has been too warm. Chutney should be thrown away if it ferments or becomes mouldy, which can result from undercooking so that there is too much water in the mixture, or from unclean jars or covers.

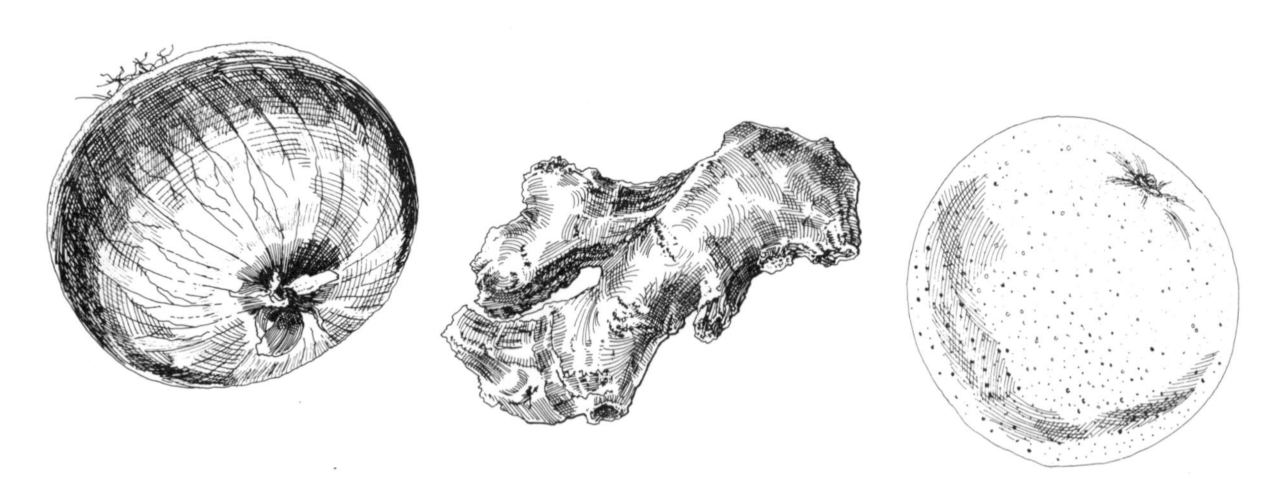

• APPLE CHUTNEY •

2.5kg/5lb apples

600ml/1 pint vinegar

500g/1lb demerara sugar

1 tbsp salt

1 tbsp ground ginger

6 chillies

250g/8oz onions

250g/8oz stoned dates

250g/8oz sultanas

Peel, core and chop the apples. Put the vinegar into a pan with the sugar, salt, ginger and chopped chillies and bring to the boil. Add the apples, chopped onions, dates and sultanas, and simmer for at least 1 hour until golden brown and thick. The apples make a lot of juice, and the chutney should not be runny. Put into hot jars, cover and seal. The chillies make the chutney very hot, and may be omitted.

APRICOT & • ORANGE CHUTNEY •

500g/1lb dried apricots

250g/8oz sugar

125g/4oz sultanas

1 garlic clove

3 peppercorns

1 orange

300ml/½ pint white vinegar

Put the apricots into a bowl, just cover with water, leave to soak overnight and then drain off the liquid. Put the fruit into a preserving pan with sugar, sultanas, chopped garlic clove, peppercorns, grated orange rind and juice. Bring to the boil. Stir in the vinegar and simmer for 45 minutes until thick. Put into hot jars, cover and seal.

• DRIED APRICOT • CHUTNEY

500g/1lb dried apricots

750g/1½lb onions

500g/1lb sugar

2 oranges

250g/8oz sultanas

900ml/1½ pints cider vinegar

2 tsp salt

2 garlic cloves

1 tsp mustard

½ tsp ground allspice

Put the apricots into a bowl and just cover with water. Leave to soak overnight, drain and chop finely. Put the apricots and chopped onions into a pan with the sugar, grated orange rind and juice. Add the sultanas, vinegar, salt, crushed garlic, mustard and allspice. Simmer gently for 1 hour until golden brown and thick, stirring occasionally to prevent sticking. Put into hot jars, cover and seal.

• AUTUMN CHUTNEY •

2kg/4lb apples

1kg/2lb pears

1.5kg/3lb red tomatoes

2kg/4lb soft brown sugar

250g/8oz sultanas

250g/8oz seedless raisins

1 litre/2 pints vinegar

1 tsp ground mace

1 tsp cayenne pepper

1 tsp ground cloves

1 tsp pepper

2 tbsp salt

1 tsp ground ginger

Peel, core and chop the apples and pears. Skin and chop the tomatoes. Put all the ingredients into a pan. Stir well and simmer for 2 hours until golden brown and thick, stirring occasionally to prevent sticking. Put into hot jars, cover and seal.

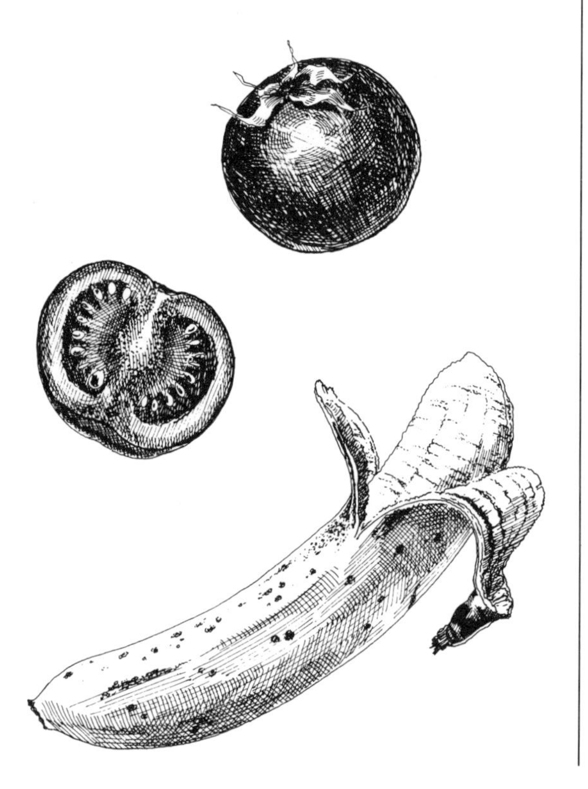

• BANANA • CHUTNEY

1kg/2lb ripe bananas

250g/8oz stoned dates

1 lemon

450ml/¾ pint vinegar

250g/8oz seedless raisins

250g/8oz demerara sugar

300ml/½ pint syrup from canned fruit

125g/4oz crystallized ginger

2 tsp salt

4 tsp curry powder

Use syrup from any canned fruit, or a mixture of varieties. Peel the bananas and cut into small pieces. Chop the dates and put into a pan with the bananas. Add the grated lemon rind and juice with the vinegar. Cover the pan and cook gently for 1½ hours. Remove lid and stir in raisins, sugar, fruit syrup, chopped ginger, salt and curry powder. Simmer for 30 minutes, stirring well, until thick. Put into hot jars, cover and seal.

• BEETROOT • CHUTNEY

1kg/2lb cooked beetroot

2 medium onions

500g/1lb apples

250g/8oz sugar

600ml/1 pint vinegar

1 tbsp lemon juice

½ tsp ground ginger

½ tsp salt

Peel the beetroot and cut into small cubes. Chop the onions and peel, core and chop the apples. Mix all the ingredients except the beetroot, and bring to the boil. Simmer for 30 minutes. Add the beetroot and simmer for 15 minutes. Put into hot jars, cover and seal.

• BLACKBERRY • CHUTNEY

2.5kg/6lb blackberries

1kg/2lb cooking apples

1kg/2lb onions

1kg/2lb soft brown sugar

1 litre/2 pints vinegar

25g/1oz salt

50g/2oz mustard

50g/2oz ground ginger

2 tsp ground mace

1 tsp cayenne pepper

Wash the blackberries, remove stems and hard berries, and put them into a preserving pan. Peel, core and chop the apples and chop the onions. Add to the blackberries with the sugar, vinegar and spices. Simmer for 1¼ hours until thick, stirring well. Put into hot jars, cover and seal.

• CHERRY • CHUTNEY

1.5kg/3lb eating cherries

250g/8oz seedless raisins

125g/4oz soft brown sugar

50g/2oz honey

300ml/½ pint vinegar

2 tsp ground mixed spice

Chop the cherries and discard the stones. Put all ingredients into a pan, heat gently and stir until the sugar has dissolved. Bring to the boil and boil for 5 minutes. Reduce the heat and simmer for 30 minutes, stirring all the time until thick. Put into hot jars, cover and seal.

• PICKLED DATES •

1kg/2lb dessert dates

white vinegar

12 peppercorns

10 cloves

small piece of cinnamon stick

1 tsp salt

Use very succulent dates and stone them carefully. Put them into preserving jars. Measure out enough vinegar to cover them, and allow the given quantities of spices to each 600ml/1 pint vinegar. Tie the spices in a piece of muslin and suspend in a pan. Add the vinegar and bring to the boil. Remove the spices and pour the hot vinegar over the dates. Cool and then seal tightly. Leave for 1 month before using.

• DAMSON • CHUTNEY

1.5kg/3lb damsons

500g/1lb apples

250g/8oz onions

250g/8oz stoned dates

250g/8oz soft brown sugar

600ml/1 pint vinegar

15g/½oz ground ginger

15g/½oz mustard

15g/½oz salt

½ tsp pepper

½ tsp ground cloves

Put the damsons into a pan and simmer in their own juice until soft, then take out the stones. Peel, core and chop the apples, and chop the onions. Put all the ingredients into a pan and simmer for 2 hours until soft and thick. Stir occasionally to prevent sticking. Put into hot jars, cover and seal.

SPICED • DAMSONS •

2kg/4lb large damsons
600ml/1 pint vinegar
750g/1½lb sugar
25g/1oz mixed pickling spice

Prick the damsons with a large needle. Put the vinegar, sugar and spice into a pan and stir over low heat until the sugar has dissolved. Add the damsons and cook until soft but not broken. Lift out the damsons with a perforated spoon and put into hot preserving jars. Boil the vinegar syrup for 5 minutes, take out the spices, and pour the liquid over the fruit. Cover tightly.

• FRESH FIG • CHUTNEY

1kg/2lb green figs
500g/1lb onions
125g/4oz crystallized ginger
600ml/1 pint vinegar
250g/8oz soft brown sugar
2 tsp salt
½ tsp pepper

Wash the figs and cut them into small pieces with the onions and ginger. Heat the vinegar, sugar, salt and pepper, stirring until the sugar has dissolved. Add the figs, onions and ginger and bring to the boil. Simmer for 45 minutes until the mixture is thick, pour into jars and cover.

• GOOSEBERRY • CHUTNEY

2kg/4lb green gooseberries
500g/1lb soft brown sugar
1 litre/2 pints vinegar
500g/1lb onions
750g/1½lb seedless raisins
125g/4oz mustard seeds
50g/2oz ground allspice
2 tsp salt

Top and tail the gooseberries. Mix the sugar with half the vinegar and boil until a syrup forms. Add the chopped onions, raisins, bruised mustard seeds, spice and salt. Boil the gooseberries in the remaining vinegar until tender. Put the two mixtures together, and simmer for 1 hour until golden brown and thick, stirring occasionally to prevent sticking. Put into hot jars, cover and seal.

GRAPE & • APPLE CHUTNEY •

1kg/2lb grapes
1kg/2lb apples
250g/8oz seedless raisins
625g/1¼lb soft brown sugar
300ml/½ pint cider vinegar
150ml/¼ pint lemon juice
½ tsp ground allspice
½ tsp ground cloves
½ tsp salt
pinch ground cinnamon
pinch paprika
½ lemon, grated rind

Cut the grapes in half and discard the seeds. Peel, core and chop the apples. Put all the ingredients into a pan and bring to the boil. Simmer for 1 hour until soft, golden brown and thick, stirring occasionally to prevent sticking. Put into hot jars, cover and seal.

• GREEN TOMATO • CHUTNEY

2.5kg/5lb green tomatoes
500g/1lb onions
15g/½oz pepper
25g/1oz salt
500g/1lb soft brown sugar
600ml/1 pint vinegar
250g/8oz seedless raisins
250g/8oz sultanas

Slice the tomatoes and chop the onions. Put them into a bowl with the pepper and salt. Mix well and leave to stand overnight. Put the sugar and vinegar into a pan and bring to the boil, then add the raisins and sultanas and bring to the boil again. Simmer for 5 minutes. Add the tomatoes and onions and continue simmering for 1 hour until golden brown and thick, stirring occasionally to prevent sticking. Put into hot jars, cover and seal.

GREEN TOMATO & • APPLE CHUTNEY •

2kg/4lb green tomatoes
500g/1lb apples
750g/1½lb onions
250g/8oz seedless raisins
500g/1lb soft brown sugar
600ml/1 pint vinegar
15g/½oz ground ginger
15g/½oz salt
12 red chillies

Cut up the tomatoes without peeling. Peel, core and chop the apples, chop the onions. Put the tomatoes, apples, onions and raisins into a pan with the sugar, vinegar, ginger and salt. Tie the chillies into a piece of muslin and suspend in the pan. Bring to the boil, stir well, and simmer for 1 hour until thick and golden brown. Remove the bag of chillies. Put into hot jars, cover and seal. The chillies make the chutney hot, and may be omitted.

• LEMON CHUTNEY •

6 large lemons
250g/8oz onions
1½ tbsp salt
water
450ml/¾ pint cider vinegar
250g/8oz light soft brown sugar
125g/4oz sultanas
1½ tbsp white mustard seeds
1 tsp ground ginger
½ tsp cayenne pepper

Do not peel the lemons, but cut them into very thin crosswise slices. Discard any pips. Chop the onions finely. Place lemon slices and onions in a large bowl and sprinkle with salt. Leave to stand in a cool place for 24 hours.

Drain the lemons and onions and place in a large pan. Just cover with water and simmer over low heat until the lemon peel is soft but not broken. Add the vinegar, sugar, sultanas and spices. Stir well and bring to the boil. Reduce heat and simmer for 1 hour, stirring frequently. Pour into hot jars, cover and seal.

• MANGO CHUTNEY •

6 ripe mangoes
300ml/½ pint cider vinegar
250g/8oz light soft brown sugar
40g/1½oz fresh ginger root
2 garlic cloves
2 tsp chilli powder
1 tsp salt

Peel the mangoes and slice them thinly. Put into a pan with the vinegar and simmer over low heat for 10 minutes. Stir in the sugar. Peel and chop the ginger, and crush the garlic. Add to the pan with the chilli powder and salt.

Bring slowly to the boil, stirring well. Reduce heat and simmer for 50 minutes, stirring occasionally. Put into hot jars, cover and seal.

MARROW & • TOMATO CHUTNEY •

1.5kg/3lb prepared vegetable marrow
75g/3oz salt
500g/1lb red tomatoes
250g/8oz apples
250g/8oz onions
1 red pepper
1 green pepper
125g/4oz stoned dates
175g/6oz seedless raisins
600ml/1 pint vinegar
15g/½oz mustard seeds
2 tsp ground ginger
2 tsp ground allspice
1 tsp ground cinnamon
1 tsp ground mace
500g/1lb sugar

Peel the marrow, remove seeds and pith, and weigh the flesh to give 1.5kg/3lb. Chop in small cubes and put in a large bowl with salt in layers. Cover and leave in a cool place for 24 hours. Drain the marrow thoroughly. Skin the tomatoes and chop roughly. Peel, core and chop the apples, and chop the onions. Remove the stems, seeds and membranes from the peppers and chop the flesh. Put all the ingredients except the marrow and sugar into a preserving pan and cook over low heat for 1½ hours. Stir in the sugar and when it has dissolved, stir in the marrow cubes. Simmer for 1 hour until the marrow is soft and most of the liquid has evaporated, stirring occasionally to prevent sticking. Put into hot jars, cover and seal.

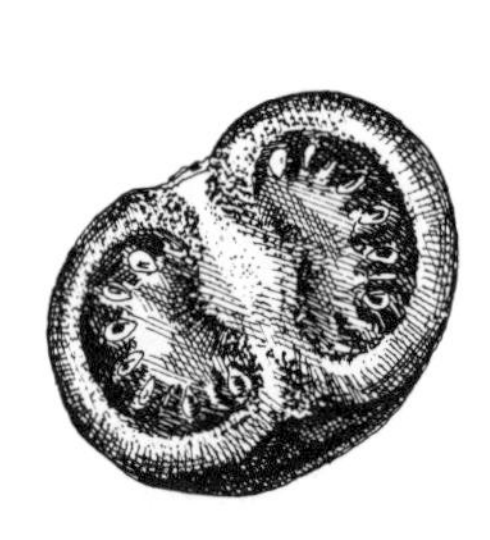

Spiced Oranges (page 130).

HOME MADE
Green & Red
Pepper Pickle
Corn Relish
HOME MADE
Strawberry Jam
Creda
HOME MADE

SPICED MELON

2kg/4lb melon

50g/2oz salt

600ml/1 pint water

500g/1lb sugar

600ml/1 pint vinegar

8 cloves

small piece of cinnamon stick

Small sweet melons are excellent for this, but not watermelon or the Ogen variety. Peel the melons and cube the flesh before weighing. Put the melon pieces into a bowl and cover with the salt and water. Leave overnight, then drain off the salt water. Cover the melon with cold water, bring to the boil and simmer until the melon cubes are transparent and tender but not broken. In another pan, boil the sugar, vinegar and spices together for 20 minutes. Strain the syrup and bring to the boil. Drain the melon and add to the syrup. Boil for 10 minutes, then pour into hot preserving jars and seal tightly.

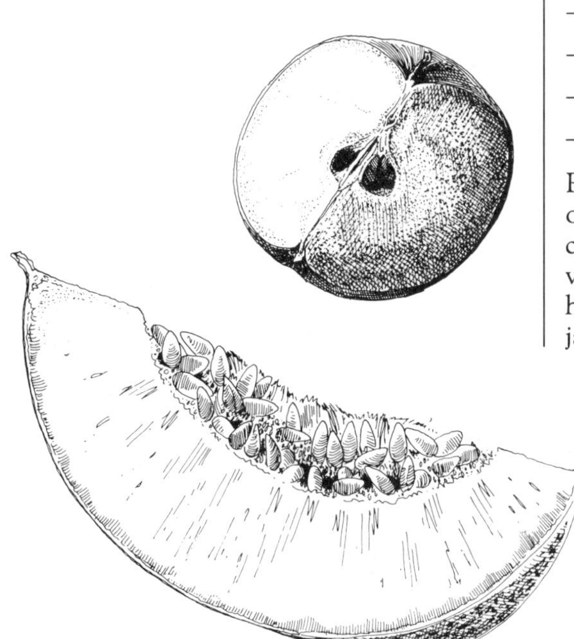

MINT CHUTNEY

375ml/¾ pint cider vinegar

500g/1lb sugar

2 tsp mustard

500g/1lb eating apples

2 medium onions

250g/8oz fresh mint leaves

75g/3oz seedless raisins

pinch salt

Heat the vinegar in a pan and then stir in the sugar and mustard until the sugar has dissolved. Stir well, take off the heat and cool slightly. Chop the apples, onions and mint very finely and put in a bowl with the raisins and salt. Pour on the vinegar mixture, mix well and pour into hot preserving jars. Seal at once.

ORANGE CHUTNEY

6 thin-skinned oranges

250g/8oz onions

500g/1lb stoned dates

600ml/1 pint vinegar

2 tsp ground ginger

2 tsp salt

Peel the oranges and discard pips. Chop the oranges and put into a pan with any orange juice, chopped onions and chopped dates. Add the vinegar, ginger and salt and stir well. Cook for 1 hour until golden brown and thick. Put into hot jars, cover and seal.

An assortment of Microwave Preserves (pages 139–143).

SPICED
• ORANGES •

6 thin-skinned oranges
450ml/¾ pint white vinegar
350g/12oz sugar
2 tsp cloves
7.5cm/3in cinnamon stick

Wipe the oranges and slice them across in thin rounds without peeling. Put into a pan with enough water to cover and simmer for 45 minutes until the peel of the fruit is tender. Drain the fruit and discard water. Put the vinegar, sugar and spices into a pan and bring to the boil. Reduce to simmering point, and add the orange rings a few at a time. Cook gently until the rind becomes transparent. Lift out the orange rings with a slotted spoon and pack into preserving jars. Boil the liquid for 5 minutes until it starts to thicken and pour over the fruit. Put one or two of the cloves into each jar and seal at once.

DEMERARA
• PEACHES •

1kg/2lb small peaches
300ml/½ pint white vinegar
500g/1lb demerara sugar
small piece of cinnamon stick
6 cloves

Wipe the peaches which should not be too ripe. Put the vinegar, sugar, cinnamon and cloves into a pan and heat slowly, stirring until the sugar has dissolved. Put in the peaches, bring to the boil, and cook over low heat until the fruit is tender but not soft. Lift out the peaches with a perforated spoon and put into hot preserving jars. Bring the liquid to the boil and pour over the peaches. Seal tightly and leave for 3 months before eating.

PICKLED
• PEACHES •

2kg/4lb small ripe peaches
1kg/2lb sugar
600ml/1 pint white vinegar
25g/1oz cloves
25g/1oz allspice berries
25g/1oz cinnamon stick

Wash and wipe the peaches. Put the sugar and vinegar in a saucepan and dissolve the sugar over gentle heat. Tie the spices into a piece of muslin, crush them, and suspend the muslin bag in the saucepan. Put in the fruit and simmer until tender but not soft. Drain the peaches with a slotted spoon and pack into preserving jars. Remove the spice bag, and boil the liquid for 10 minutes until syrupy. Cover the fruit and seal at once.

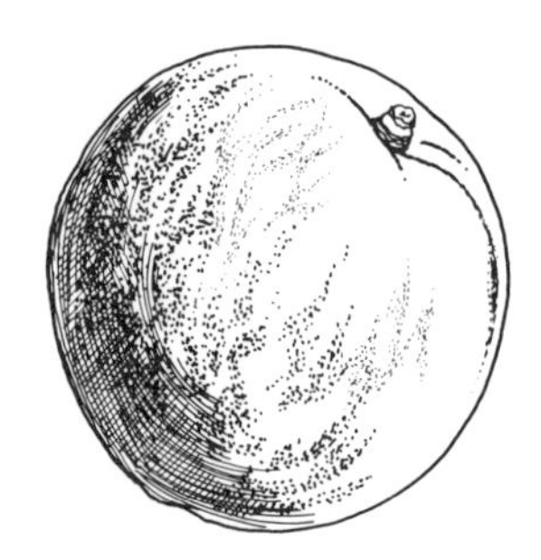

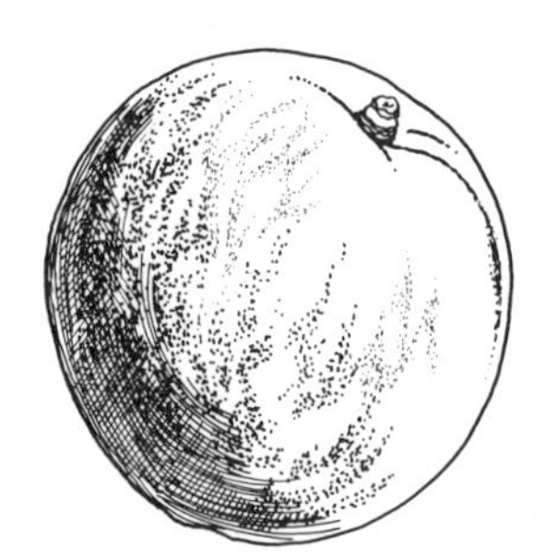

• PEAR CHUTNEY •

2kg/4lb eating pears

750g/1½lb soft brown sugar

500g/1lb seedless raisins

2 oranges

300ml/½ pint white vinegar

1 tsp ground cloves

1 tsp ground cinnamon

1 tsp ground allspice

Peel and core pears and chop them. Put the pears, sugar, raisins, grated orange rind and juice into a pan with the vinegar and spices. Bring to the boil and simmer for 1 hour until thick and golden brown, stirring occasionally to prevent sticking. Put into hot jars, cover and seal.

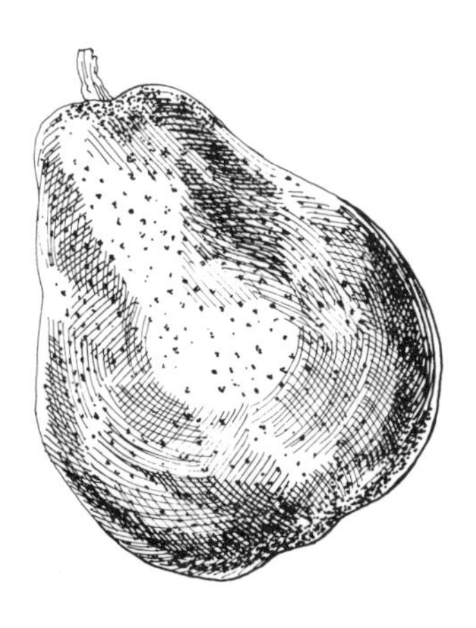

PICKLED • PEARS •

2kg/4lb cooking pears

1kg/2lb sugar

600ml/1 pint white vinegar

25g/1oz cloves

25g/1oz allspice berries

25g/1oz root ginger

25g/1oz cinnamon stick

½ lemon, rind

Peel and core the pears and cut into quarters. Put the sugar and vinegar in a pan and dissolve the sugar over gentle heat. Put the spices and lemon rind into a piece of muslin, crush them, and suspend the muslin bag in the saucepan. Put the pears into the vinegar and simmer until tender. Lift out the pears with a slotted spoon and pack into preserving jars. Remove the spice bag, and boil the liquid for 10 minutes until syrupy. Cover the fruit and seal at once.

• PINEAPPLE • CHUTNEY

1 medium pineapple

250g/8oz sugar

150ml/¼ pint cider vinegar

1 tsp ground cloves

1 tsp ground cinnamon

1 tsp ground ginger

1 tbsp curry powder

Peel and core the pineapple and cut flesh in small pieces. Put any pineapple juice with the sugar and vinegar into a pan and bring to the boil. Add the spices and simmer for 5 minutes. Add the pineapple and cook for 10 minutes on low heat, stirring well. Put into hot jars, cover and seal.

• PLUM CHUTNEY •

1.5kg/3lb plums

500g/1lb carrots

25g/1oz garlic

25g/1oz chillies

600ml/1 pint vinegar

500g/1lb seedless raisins

500g/1lb soft brown sugar

25g/1oz ground ginger

40g/1¼oz salt

A mixture of plums can be used for this chutney. Cut up the plums and discard the stones. Mince the carrots and chop the garlic and chillies. Put plums, carrots and vinegar into a pan and simmer until soft. Add the remaining ingredients and simmer for 1 hour until thick and soft. Stir occasionally to prevent sticking. Put into hot jars, cover and seal.

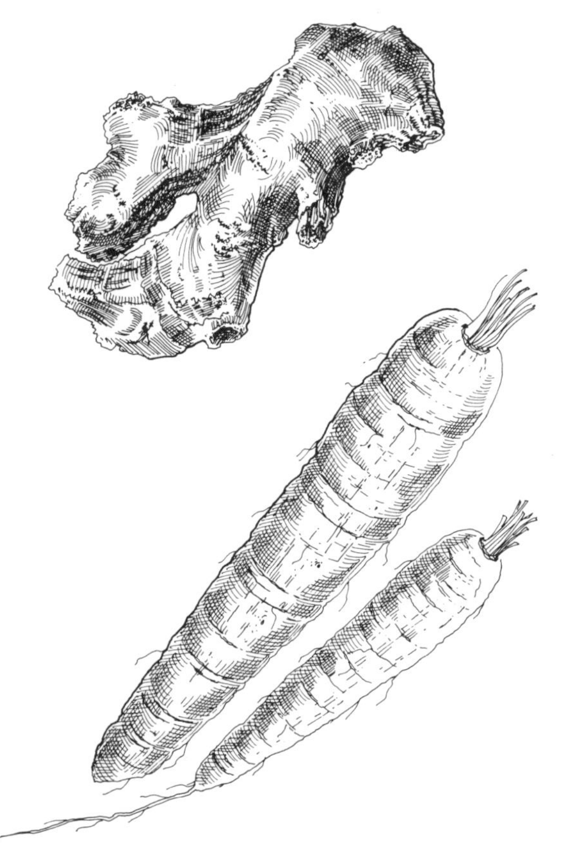

PICKLED • PLUMS •

1kg/2lb plums

600ml/1 pint vinegar

750g/1½lb sugar

2.5cm/1in cinnamon stick

1 tsp cloves

10 allspice berries

blade of mace

Small black eating plums are best for this. Wipe them and prick 4 times each with a needle, and put into a bowl. Put the vinegar, sugar and spices into a pan and boil for 10 minutes. Pour over the plums and leave overnight. Drain off the liquid and boil for 10 minutes. Pour over the fruit again and leave for 12 hours. Put the plums into a pan with the liquid and bring to the boil. Remove the spices. Lift out the plums with a slotted spoon and pack into preserving jars. Boil the syrup, pour over the plums and seal at once.

• RED TOMATO • & CHUTNEY •

500g/1lb red tomatoes

125g/4oz apples

250g/8oz onions

500g/1lb seedless raisins

125g/4oz soft brown sugar

300ml/½ pint vinegar

2 tsp salt

2 tsp ground ginger

pinch cayenne pepper

Skin the tomatoes and chop the flesh. Peel, core and chop the apples, and chop the onions. Put all the ingredients into a pan. Stir well and simmer for 1 hour until thick and golden brown. Put into hot jars, cover and seal.

• RHUBARB • CHUTNEY

1kg/2lb rhubarb

250g/8oz onions

750g/1½lb soft brown sugar

250g/8oz sultanas

600ml/1 pint vinegar

15g/½oz mustard

1 tsp ground mixed spice

1 tsp pepper

1 tsp ground ginger

1 tsp salt

pinch cayenne pepper

Wipe the rhubarb and cut into short lengths. Mix with the chopped onions in a pan and add all the ingredients. Stir well and simmer for 1 hour until soft, golden brown and thick, stirring occasionally to prevent sticking. Put into hot jars, cover and seal.

• RUNNER BEAN • CHUTNEY

1kg/2lb runner beans

750g/1½lb onions

900ml/1½ pints vinegar

25g/1oz cornflour

15g/½oz turmeric

25g/1oz mustard

500g/1lb demerara sugar

Slice the beans in medium shreds and chop the onions. Cook the beans in salted water until tender and drain well. Cook the onions in one-third of the vinegar until tender. Mix the cornflour, turmeric and mustard with a little vinegar to a smooth paste, and then add the remaining vinegar and beans and cook for 10 minutes. Add the sugar, onions and vinegar in which they have been cooked and boil for 15 minutes. Pour into preserving jars and seal.

• SAUCES •

The same equipment as for pickles and chutney can be used for sauces, with the addition of a large hair or nylon sieve. An easy sauce can be made from any chutney recipe if the mixture is rubbed through a sieve about two-thirds of the way through the cooking time and then cooked again, but sauces must be sterilized in their bottles. Ketchup is thinner than sauce and a jelly bag is needed for straining.

Cut-up fruit and vegetables need to be simmered in vinegar until soft, and then the sieved purée is sweetened and spiced and cooked gently until smooth and creamy with a bright colour and distinct flavour. Ketchup will be thin and almost clear. Both can be bottled in sterilised sauce bottles with vinegar-proof screw tops, but the filled bottles must be sterilised as well. Sauces should be stored in a cool dark place so that they do not discolour. If sauce does not keep, it may be because bottles and covers were not clean, or correctly sterilised. If sauce ferments or is mouldy, it must be thrown away, and this may have occurred because covers were not airtight or the storage place was too warm. If the mixture is not cooked long enough, there may be too much water left in, and this will also cause deterioration. Sometimes sauce had a slightly

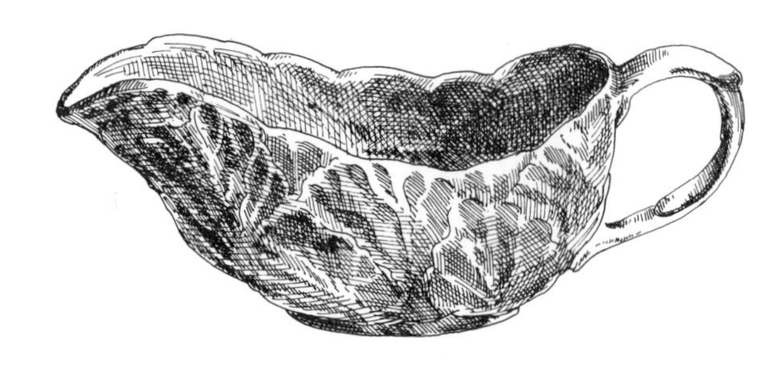

watery appearance and separates during storage, which probably means it has not been cooked quite long enough, and it is best shaken well to restore smoothness.

• Sterilisation

Fill clean bottles with hot sauce and screw on clean tops tightly. Release tops a half-turn and put bottles in a large pan on a rack or thick pad of newspaper. Cover the bottles with hot but not boiling water and bring to 54°C/150°F. Keep the water simmering at this temperature for 30 minutes. Remove bottles, and screw on tops tightly at once.

• BLACKBERRY • SAUCE

3kg/6lb blackberries

water

600ml/1 pint vinegar

1 tsp salt

25g/1oz sugar

1 tsp mustard

½ tsp ground cloves

½ tsp ground nutmeg

½ tsp ground cinnamon

Clean blackberries and remove stems and unripe berries. Put into a pan with just enough water to cover and simmer until the berries are soft. Strain through a sieve, and put the purée into a pan with the remaining ingredients. Simmer for 15 minutes, put into hot bottles and seal. Sterilise for 30 minutes in a water bath.

• CRANBERRY • SAUCE

1kg/2lb cranberries

250g/8oz onions

250ml/8fl. oz water

250g/8oz sugar

300ml/½ pint vinegar

½ tsp ground cloves

½ tsp ground cinnamon

½ tsp ground allspice

½ tsp pepper

1 tbsp salt

Put the cranberries, chopped onions and water into a pan, cover and simmer for 30 minutes until the cranberries are soft. Put through a sieve and return purée to pan. Stir in the sugar, vinegar, spices and salt, and simmer for 20 minutes. Put into hot bottles, seal and sterilise for 30 minutes in a water bath.

• FRUIT SAUCE •

3.5kg/7lb tomatoes
3kg/6lb cooking apples
1kg/2lb onions
500g/1lb stoned dates
1.5 litres/3 pints white vinegar
1 tbsp salt
1 tsp ground mixed spice
25g/1oz mustard seeds
15g/½oz cloves
½ tsp cayenne pepper
2 tsp ground ginger
3 red chillies
2 blades mace
1.5kg/3lb seedless raisins
1.5kg/3lb soft brown sugar

Use either red or green tomatoes or a mixture. Slice the tomatoes. Peel, core and chop the apples. Chop the onions and dates. Put all the ingredients except the sugar into a pan and simmer for 1 hour until soft. Put through a sieve and return to pan with the sugar. Stir over low heat until the sugar has dissolved, bring to the boil, and simmer for 45 minutes until thick. Put into hot bottles and seal. Sterilise for 30 minutes in a water bath.

• GOOSEBERRY • SAUCE

1kg/2lb gooseberries
350g/12oz seedless raisins
500g/1lb onions
250g/8oz soft brown sugar
600ml/1 pint vinegar
1 tsp powdered mustard
1 tbsp ground ginger
2 tbsp salt
pinch cayenne pepper
2 tsp turmeric

Top and tail the gooseberries, and mince them with the raisins and onions. Put into a pan with the remaining ingredients, bring slowly to the boil and simmer for 45 minutes. Sieve and return purée to the pan. Simmer for 15 minutes, stirring occasionally, put into hot bottles and seal. Sterilise for 30 minutes in a water bath.

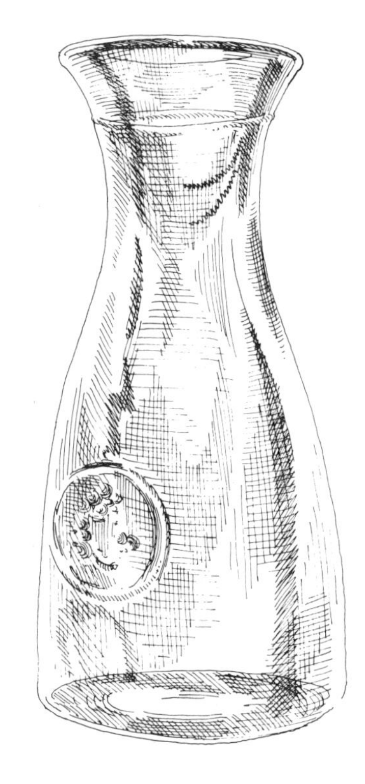

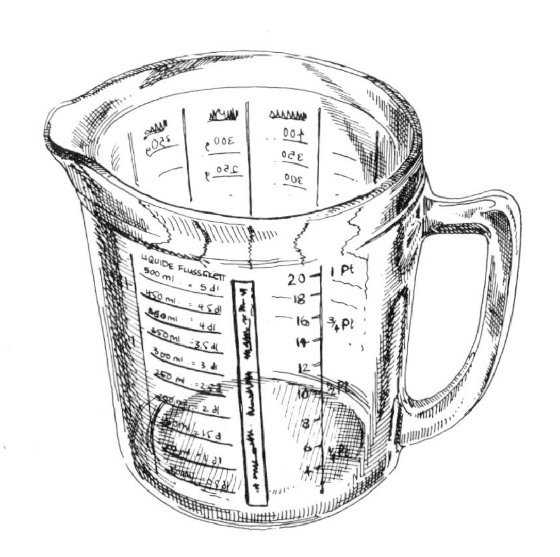

• GREEN TOMATO • SAUCE

1kg/2lb green tomatoes

125g/4oz shallots

15g/½oz celery seed

1 lemon

2 tsp turmeric

1 tsp ground allspice

½ tsp ground ginger

½ tsp ground cinnamon

250g/8oz sugar

300ml/½ pint vinegar

Cut up the tomatoes and put in a pan with chopped shallots and celery seed. Chop the lemon without peeling and add to the pan. Cover and simmer very gently for 40 minutes until soft. Put through a sieve and return to pan. Add remaining spices, sugar and vinegar, bring to the boil and simmer for 45 minutes until thick and smooth. Put into hot bottles and seal. Sterilise for 30 minutes in a water bath.

• MINT SAUCE •

150ml/¼ pint mint

300ml/½ pint vinegar

150g/6oz demerara sugar

Chop the mint finely and press it down in a measuring jug to get the correct amount. Boil the vinegar and sugar together for 2 minutes, stirring well so that the sugar dissolves. Add the chopped mint, stir well and leave until cold. Put into screw-top jars with vinegar-proof lids. For use, stir well, take out required amount and add a little extra vinegar.

• MUSHROOM • KETCHUP

1kg/2lb field mushrooms

75g/3oz salt

600ml/1 pint vinegar

1 tsp allspice berries

1 tsp black peppercorns

1 small piece root ginger

6 blades mace

4 cloves

2.5cm/1in cinnamon stick

Trim the base of each mushroom stalk and break the mushrooms into small pieces. Arrange in layers with salt in a bowl. Cover with a cloth and leave in a cool place for 5 days, stirring once or twice a day. Cover the bowl with a piece of foil and put into the oven at 150°C/300°F/gas mark 2 for 1½ hours. Strain through a jelly bag and leave to drip for an hour until all the liquid has dripped through. Put the liquid into a pan with the vinegar and spices, and simmer gently until the liquid is reduced by half. Strain through muslin, bring to the boil again, put into hot bottles, cover and seal. Sterilise for 30 minutes in a water bath.

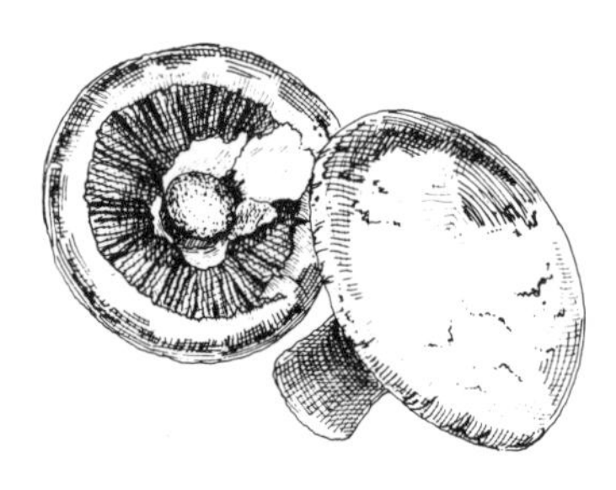

• PLUM SAUCE •

1kg/2lb plums
250g/8oz sugar
600ml/1 pint vinegar
1 tsp salt
1 tsp ground ginger
½ tsp cayenne pepper
8 cloves

A mixture of plums can be used, with some damsons if available. Cut up the plums and put with the stones and all the ingredients into a pan. Stir well, bring to the boil, and simmer for 30 minutes. Sieve and return purée to the pan. Simmer for 30 minutes, stirring occasionally. Put into hot bottles and seal. Sterilise for 30 minutes in a water bath.

• RASPBERRY • SAUCE

2kg/4lb raspberries
600ml/1 pint cider vinegar
1 tsp salt
1 tsp mustard
½ tsp ground mixed spice
350g/12oz sugar

Put ripe fruit into a pan with vinegar and simmer for 15 minutes. Strain through a sieve, pressing out the juice, and return liquid to pan. Add the salt, mustard and spice, and simmer for 30 minutes. Strain again through a sieve and stir in the sugar. Simmer for 30 minutes, put into hot bottles and seal. Sterilise for 30 minutes in a water bath.

• TOMATO SAUCE •

2kg/4lb red tomatoes
4 large onions
500g/1lb demerara sugar
600ml/1 pint vinegar
50g/2oz peppercorns
25g/1oz salt
15g/½oz cloves
2 tsp cayenne pepper

Slice the tomatoes and onions, and put into a pan with the other ingredients. Stir well and simmer for 2 hours until thick and soft, stirring occasionally. Sieve and return purée to pan. Bring to the boil and boil for 5 minutes. Pour into hot bottles and seal. Sterilise for 30 minutes in a water bath.

• WALNUT KETCHUP •

80 green walnuts
1.5 litres/3 pints vinegar
250g/8oz chopped onions
175g/6oz salt
15g/½oz peppercorns
15g/½oz allspice berries
12 cloves
6 blades mace

Pick the walnuts before the shells have formed inside the green casing. Split them and crush them in a large bowl, or mince coarsely. Put all the other ingredients into a saucepan and bring to boiling point. Pour over the nuts. Cover and leave in a cold place for 14 days, stirring once or twice each day. Strain the liquid through a jelly bag for 2–3 hours. Put the liquid into a pan, bring to the boil, and simmer for 1 hour. Put into hot bottles, seal and sterilise for 30 minutes in a water bath.

• HERB VINEGAR •

fresh herbs

white wine vinegar

Use tarragon, basil, burnet, marjoram or mint, or a mixture of herbs. Pick them as soon as the flower-buds appear, but before they open. Pack a jar full of leaves and fill with vinegar. Seal and leave for at least 2 weeks. Strain and pour into bottles. Use for salad dressings and mayonnaise.

• GARLIC VINEGAR •

8 garlic cloves

salt

600ml/1 pint vinegar

Crush the garlic cloves with a little salt. Bring vinegar to boiling point and pour over the garlic. Put into a jar, cool, cover and leave for at least 2 weeks. Strain and bottle. Use for salad dressings and mayonnaise.

MICROWAVE PRESERVES

It is quick and easy to make both sweet and vinegar preserves in a microwave oven. There is no smell and no steam, and only an oven-glass bowl to wash. The colour and flavour of these preserves is superb. Only small quantities may be made (e.g. 1.5kg/3lb) but this is no problem for small households, or when fruit is expensive. Indeed, such small quantities are ideal when experimenting with new fruits or chutneys. If a larger quantity is needed, recipes should not be doubled, but should be prepared in small batches.

The only equipment needed is a large bowl which will fit easily into the microwave oven, and oven-glass is preferable as it is easy to check the progress of cooking. A wooden spoon is needed for stirring. A sugar thermometer must not be used in the oven, but obviously may be used outside the oven for testing setting point. Alternatively, you can test preserves with any of the conventional setting tests (see page 83).

Ingredients and preparation are the same as for conventional cooking, but less liquid is used in microwave recipes as there is not much evaporation in the oven. When preparing marmalade, it is important to shred peel finely or it will not soften as much as by using a conventional method.

It is important to use the specified cooking times, and not to be tempted to add on a few minutes, as sugar mixtures burn quickly in a microwave oven.

To speed up cooking, the sugar may be heated before adding to the other ingredients. Just microwave the opened bag for 2 minutes on full power. Make sure that the sugar is completely dissolved before final cooking or it may burn in the bottom of the bowl and discolour and flavour the preserve.

The testing, potting and covering of microwave preserves is exactly the same as for conventional preparation. Jars may be sterilised by quarter-filling them with water and heating on full power until the water is boiling. Remove carefully with oven gloves, and drain upside down on absorbent paper.

Important. Sugar mixtures are very hot indeed. Oven gloves must be used to lift the bowl from the microwave oven, and great care must be taken.

MICROWAVE
• SPICED APPLE • BUTTER

1kg/2lb Bramley apples

150ml/¼ pint dry cider

1 lemon, thinly peeled rind

water

500g/1lb light soft brown sugar

¼ tsp ground cinnamon

¼ tsp ground cloves

¼ tsp ground ginger

Do not peel or core the apples, but chop them roughly. Put into a large bowl with the cider, lemon rind and 150ml/¼ pint water. Cover and cook on full power for 15 minutes. Discard the lemon rind, and sieve the apples.

Return the apple purée to the bowl. Stir in the sugar and spices and cook on full power for 15 minutes, stirring frequently. Continue cooking for 20 minutes until thick and creamy. Pour into warm jars, cover and seal.

MICROWAVE
• APPLE CHUTNEY •

1kg/2lb cooking apples

750g/1½lb cooking dates

500g/1lb onions

175g/6oz black treacle

125g/4oz dark soft brown sugar

1 tsp salt

1 tsp ground allspice

1 tsp cayenne pepper

600ml/1 pint vinegar

Peel and core the apples and chop them roughly. Chop the dates, and chop the onions finely. Put all the ingredients into a large bowl. Cover and cook on full power for 10 minutes, stirring occasionally.

Continue cooking on full power for 45 minutes, stirring two or three times. Cool for 5 minutes, stirring once or twice. Put into warm jars, cover and seal.

MICROWAVE
CUCUMBER &
• ONION PICKLE •

2 large cucumbers

2 medium onions

25g/1oz salt

300ml/½ pint white vinegar

75g/3oz sugar

½ tsp celery seeds

½ tsp mustard seeds

Do not peel the cucumbers, but cut them into ¾cm/½in cubes. Chop the onions finely. Mix together in a large bowl and sprinkle with the salt. Leave to stand for 2 hours.

Drain away liquid and rinse the cucumbers and onions well in cold water. Put the vinegar, sugar, celery and mustard seeds into a bowl and microwave for 6 minutes, stirring occasionally. Stir in the cucumber and onions and microwave for 2 minutes. Put into warm jars, cover and seal.

MICROWAVE
• KIWI FRUIT JAM •

6 large kiwi fruit

1 lime, juice

500g/1lb sugar

Peel the kiwi fruit and chop them roughly. Put into a bowl with the lime juice. Cover and cook on full power for 10 minutes until soft. Stir in the sugar until dissolved. Cook uncovered for 12 minutes, stirring occasionally. Pour into hot jars, cover and seal.

MICROWAVE
• LEMON CURD •

125g/4oz unsalted butter

3 large lemons

250g/8oz sugar

3 eggs

1 egg yolk

Put the butter into a bowl. Grate the lemon rind and keep on one side. Squeeze out the lemon juice and add to the butter. Heat on full power for 3 minutes. Stir in the sugar and cook for 2 minutes. Beat together the eggs and yolk and stir into the butter mixture. Stir in the lemon rind. Cook on 30 per cent power for 12 minutes, stirring occasionally. Pour into warm jars, cover and seal.

MICROWAVE
• LEMON •
MARMALADE

6 large lemons

water

1kg/2lb sugar

Peel the lemons thinly and shred the peel finely, keeping on one side. Squeeze the juice into a large bowl. Chop the pith and flesh roughly and add to the bowl. Tie pips into a piece of muslin and add to the bowl. Pour on 900ml/1½ pints boiling water. Cook on full power for 20 minutes.

Strain into another large bowl, pressing well, and discard the pulp and bag of pips. Add the peel to the liquid and cook on full power for 12 minutes. Stir in the sugar until completely dissolved. Half-cover the bowl and cook for 15 minutes. Cool for 30 minutes, stirring occasionally. Pour into warm jars, cover and seal.

MICROWAVE

• PICCALILLI •

250g/8oz green tomatoes

250g/8oz cucumber

250g/8oz dwarf beans

250g/8oz celery

2 courgettes

2 medium onions

250g/8oz cauliflower florets

50g/2oz salt

water

600ml/1 pint vinegar

pinch of ground cinnamon

pinch of chilli powder

75g/3oz sugar

25g/1oz cornflour

1 tsp mustard powder

1 tsp ground ginger

¼ tsp turmeric

Dip the tomatoes in boiling water and skin them. Chop the flesh roughly into a large bowl. Chop the cucumber, beans and celery and slice the courgettes and onions. Mix with the tomatoes and cauliflower florets. Sprinkle with salt and cover with cold water. Cover and leave for 24 hours. Drain and rinse the vegetables and put into a large bowl.

Mix the vinegar, cinnamon and chilli powder in a bowl and microwave for 5 minutes. In another bowl, mix the sugar, cornflour, mustard, ginger and turmeric. Add a little of the vinegar and mix to a paste. Stir in the rest of the vinegar. Microwave for 3 minutes, stirring twice.

Pour this mixture over the vegetables and stir well. Cover and cook for 12 minutes, stirring frequently. Cool for 5 minutes, stir well and put into warm jars. Cover and seal.

MICROWAVE
• PLUM CONSERVE •

500g/1lb plums
1 thin-skinned orange
500g/1lb sugar
125g/4oz sultanas

Cut the plums in half and discard the stones. Slice the plums roughly and put into a large bowl. Mince the orange (including the peel) or chop finely, and add to the plums. Cover and cook on full power for 15 minutes. Stir in the sugar and sultanas, and cook uncovered for 15 minutes, stirring occasionally so that the sugar dissolves. Continue cooking without stirring for 3 minutes. Pour into warm jars, cover and seal.

MICROWAVE
• RASPBERRY JAM •

500g/1lb raspberries
2 tbsp lemon juice
500g/1lb sugar

Put the fruit, lemon juice and sugar into a large bowl and stir well. Microwave on full power for 5 minutes, stirring occasionally, until the sugar has dissolved. Cook on full power for 12 minutes. Stir well, pour into warm jars, cover and seal.

MICROWAVE
• SEVILLE ORANGE •
MARMALADE

1kg/2lb Seville oranges
1 large lemon
water
2kg/4lb sugar

Cut the oranges and lemon in half and squeeze the juice into a large bowl. Remove the pips and tie them into a piece of muslin. Shred the peel finely and put into the bowl with the pips and 300ml/ $3\frac{1}{2}$ pint boiling water. Leave to soak for 1 hour.

Add 900ml/$1\frac{1}{2}$ pints boiling water. Cover and cook on full power for 20 minutes until the peel is tender. Add the sugar and stir until it has dissolved completely. Cook on full power for 45 minutes, stirring occasionally. Leave to stand for 30 minutes and stir well before pouring into warm jars. Cover and seal.

MICROWAVE
• STRAWBERRY JAM •

1.5kg/3lb strawberries
1 lemon, juice
1.5kg/3lb sugar

Put the strawberries and lemon juice into a large bowl and cook on full power until soft. Stir in the sugar and continue cooking for 40 minutes, stirring occasionally. Leave to stand for 10 minutes, stirring often to distribute the fruit. Pour into warm jars, cover and seal.

MICROWAVE • THREE FRUIT • MARMALADE

1 large grapefruit
3 sweet oranges
2 lemons
water
1kg/2lb sugar

Peel the fruit, avoiding white pith, and shred thinly. Put into a large bowl and reserve. Remove excess pith from the fruit, and then chop the fruit finely (using a food processor if liked). Put into a bowl. Tie pips and excess pith into a piece of muslin and put into the bowl. Pour on 900ml/1½ pints boiling water.

Cook on full power for 20 minutes. Strain over the peel, pressing well and cook on full power for 12 minutes. Stir in the sugar until completely dissolved. Half-cover the bowl and cook for 15 minutes. Leave to stand for 30 minutes, stirring occasionally. Pour into warm jars, cover and seal.

MICROWAVE • TOMATO CHUTNEY •

1.5kg/3lb ripe tomatoes
125g/4oz onions
2 tsp salt
1 tsp paprika
150ml/¼ pint vinegar
175g/6oz sugar

Dip the tomatoes in boiling water and skin them. Chop the flesh roughly. Chop the onions finely. Put the tomatoes and onions into a bowl with salt and paprika. Cook on full power for 30 minutes, stirring occasionally.

Stir in the vinegar and continue cooking for 15 minutes. Stir in the sugar until dissolved, and cook for 30 minutes. Cool for 5 minutes, stirring occasionally. Pour into warm jars, cover and seal.

CANDYING & CRYSTALLISING

A great variety of candied fruit and crystallised flowers may be prepared at home, which will save considerably on the purchase of shop products, and they can be used as sweetmeats, or in the making of cakes and puddings.

• CANDIED FRUIT •

Candied or glacé fruit can be made from both fresh and canned fruit. Cherries, grapes, oranges, pears, pineapple and stone fruits are suitable for use when fresh. Many canned fruits are easier to process than their fresh equivalent, and apricots, pineapple, pears, mandarin oranges and lychees are particularly good.

Choose good quality fresh fruit, or a good brand of canned fruit in which the fruit is firm and in large pieces. Cook fresh fruit gently in water until just tender, or drain the syrup from canned fruit.

Fresh fruit. Use a syrup made from 300ml/½ pint water and 175g/6oz sugar to each 500g/1lb fruit. For *canned fruit*, allow 300ml/½ pint canning

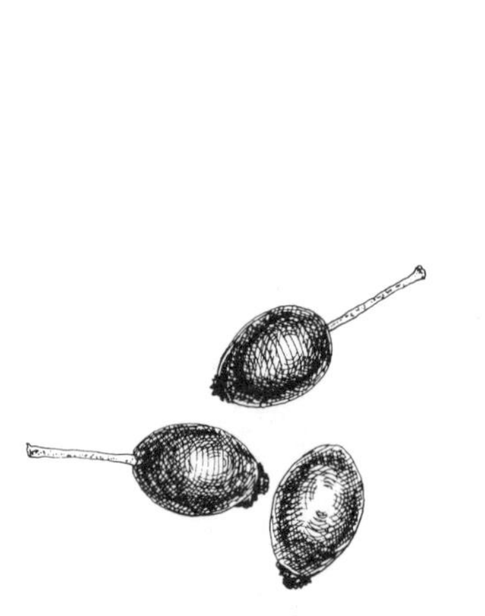

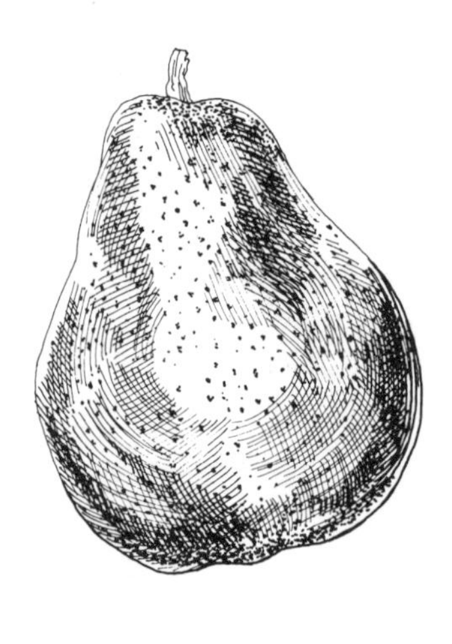

Candied Fruit (pages 145–6) and crystallised flower petals (pages 147–8).

WEST
NORWICH
Gum Arabic

syrup to each 500g/1lb fruit, making up the liquid with water if necessary – only one kind of syrup should be used, and not a mixture from cans, even if a quantity of mixed fruit is being prepared at the same time.

Heat the syrup and use it to cover the fruit completely in a large bowl. Put a plate over the fruit to keep it under the syrup, and leave to stand for 24 hours. *Day one:* Drain off the syrup, add 50g/2oz sugar and dissolve. Boil and pour over the fruit and leave 24 hours. *Days two and three:* Repeat the process each day, adding 50g/2oz sugar each time. *Day four:* Leave the fruit in the syrup. *Day five:* Add 75g/3oz sugar and boil the fruit in syrup for 4 minutes, then leave for 2 days. *Day eight:* Add 75g/3oz sugar and boil the fruit in syrup for 4 minutes, then leave the fruit to soak for 4 days.

At the end of this time, drain off the syrup and put the fruit on a wire cake rack to drain. Leave in a warm airing cupboard for about 3 days, or finish in an oven which is cooling after baking. The oven temperature should not be above 50°C/120°F/gas mark ¼ or you will risk burning the fruit and ruining the flavours.

To get a professional finish on the fruit, there are two ways of completing the processing. The most simple way is to dip the fruit quickly in boiling water and then roll each piece in a little caster sugar. A better way is to make up a solution of 250g/8oz sugar in 65ml/2½fl. oz water and bring this to the boil. Then dip the fruit in boiling water for 20 seconds and into the boiling syrup mixture. Put on a wire rack to dry, and then store in boxes with waxed paper between the layers and keep in a cool, dark, dry place. Waxed paper from cereal packets can be used for this purpose.

Good candied fruit should be firm outside and succulent inside, with a bright colour and sweet, true fruit flavour.

CANDIED • ANGELICA •

fresh angelica stems

8g/¼oz salt

2 litres/4 pints water

sugar

Cut stems from young plants in April, and cut them into 7½cm/3in pieces. Cover with boiling brine made from 8g/¼oz salt to 2 litres/4 pints water. Leave for 10 minutes and drain. Rinse angelica in cold water. Put into boiling water and boil for 7 minutes until tender. Drain and scrape off the outer skin. Weigh the stems and allow an equal weight of sugar. For each 500g/1lb sugar and 500g/1lb angelica allow 600ml/1 pint water. Dissolve the sugar in the water, bring to the boil and pour over the stems. Add 500g/1lb sugar to each original 500g/1lb angelica again, and bring to the boil. Pour syrup over stems. Repeat this process for a total of 4 days. Leave stems to soak in syrup for 2 weeks. Drain the stems and leave on paper on a cooker rack to dry slowly.

Home-made Pot Cheese (page 156).

CANDIED • BLACK CHERRIES •

1kg/2lb firm black cherries

1kg/2lb sugar

600ml/1 pint water

The cherries should be weighed after stoning. Dissolve the sugar in the water over a low heat without boiling. When the syrup is clear, put in the cherries. Simmer very gently until the cherries are almost transparent. Drain the fruit and put on flat trays. Dry thoroughly in the sun or in a very cool oven with the door slightly open. Dust with icing sugar containing a pinch of bicarbonate of soda and store in a box with waxed paper between the layers.

CANDIED • CHESTNUTS •

1kg/2lb chestnuts

1kg/2lb sugar

600ml/1 pint water

1 vanilla pod

Remove outer casings from chestnuts and boil the nuts in some water for 8 minutes. Remove the inner skin. Make a syrup with the sugar, water and vanilla pod, and when it is thick, put in the chestnuts and boil gently for 10 minutes. Remove the vanilla pod and pour nuts and syrup into a bowl and leave overnight. Reheat the syrup and chestnuts and boil for 1 minute, then pour back into the bowl and leave for 24 hours. Repeat the process three more times, until the syrup has been absorbed. Put the chestnuts on a wire rack covered with paper and dry them slowly in an open oven. Store in a wooden box lined with greaseproof paper and store in a cool dry place. Chestnuts dry more quickly than other candied fruits and lose their flavour, so they should be eaten quickly.

CANDIED • GRAPEFRUIT PEEL •

thin-skinned grapefruit

sugar

Take the rinds from the grapefruit and cut into 12mm/½in strips. Cover with cold water, bring to the boil and simmer for 5 minutes. Strain and return to the pan. Repeat process three times adding fresh water each time. In the final process, simmer the rinds until tender, strain and cover with cold water. Drain and weigh the rinds and weigh an equal quantity of sugar. Use the drained grapefruit liquid and make a syrup with the sugar. Simmer until syrup is clear, then add peel and boil until the syrup is thick. Drain the grapefruit peel on a wire rack in a very cool oven with the door open. Roll peel in granulated sugar with a pinch of bicarbonate of soda. The peel can also be dipped in melted plain chocolate to serve as a sweetmeat.

• CANDIED ORANGE • & LEMON PEEL

oranges

lemons

sugar

Remove the peel carefully from the fruit, if possible in quarters. Put the peel into a pan with enough water to cover and simmer for 1½ hours, adding more water if necessary. Add 50g/2oz sugar for each fruit used and stir until dissolved. Bring to the boil, then put aside without a lid until the next day. On the next day, bring to the boil and simmer for 5 minutes. On the following day, simmer until the peel has absorbed nearly all the syrup. Drain the peel and put on paper on a wire cake rack. Cover with greaseproof paper and dry slowly. A little syrup can be poured into the hollow of the peel pieces.

• CRYSTALLISED FLOWERS •

Crystallised flowers may be made to use as sweetmeats or decorations, and if prepared by the long-term method, they will keep for several months and retain natural colours. Flowers from bulbs should not be eaten, and the best flowers to use are primroses, violets, polyanthus, roses, carnation petals, mimosa, cowslips, sweet peas, fruit blossoms, borage. For the beginner, flowers of the primrose or polyanthus, or rose petals, are the easiest to handle as they have firm and well defined petals.

For short-term use, simply brush petals with beaten egg white and dip in caster sugar, then dry on a wire cake rack for about 24 hours. Use a small paintbrush (one from a child's paint box is about the right size) and handle the flowers carefully so they do not bruise.

For long-term storage, the flowers must be crystallised in a solution of gum arabic and rose- or orange-flower water. Crystals (not powder) of gum arabic can be bought from chemists, but may have to be ordered. Allow 3 tsp crystals to 3 tbsp rose-water or orange-flower water and put them into a screw-top jar. Leave for 3 days, shaking occasionally until the mixture is a sticky glue. Use a small, soft paintbrush to paint the flowers. For large flowers such as roses, it will be necessary to take the flowers apart, process the petals, and then reassemble the flowers for use. Paint the flowers very carefully so that petals are completely coated, as bare spots will shrivel and not keep. Colours should remain natural and delicate, but a very little vegetable food colouring may be added if the colours need brightening. When each flower is painted with gum arabic solution, dust it with caster sugar. Use a small teaspoon to sprinkle on the sugar, rather than dipping the flower into the sugar which will make the resulting crystallisation rather clogged and heavy. Put the completed flowers on a wire cake rack and leave in a warm dry place for 24 hours until crisp and dry before storage in a screw-top jar or a box.

CRYSTALLISED • MINT LEAVES •

fresh mint leaves

egg white

granulated sugar

Use fresh green mint with well-shaped leaves. Beat egg white stiffly and coat both sides of the leaves. Coat with sugar and put on a wire rack covered with waxed paper until dry. Store in a tin between layers of waxed paper.

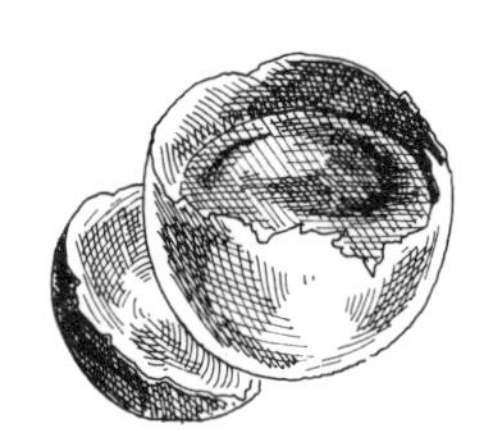

• CANDIED ROSES • & VIOLETS

1 cup small rosebuds or 2 cups violets

100ml/4fl. oz water

250g/8oz granulated sugar

The flowers should be gathered early when the dew has just dried. Bring the water to the boil and remove from heat. Stir in the sugar until dissolved. Remove stems from the flowers, and lightly wash and drain them without bruising in a colander. Put the syrup back on the heat and stir in the flowers. Cook gently to 115°C/240°F (soft ball). Take off the heat and stir with a wooden spoon until the syrup begins to granulate to the texture of coarse meal. Pour the contents of the pan into a colander and shake off the extra sugar as the flowers cool. Store in jars with the lids sealed with sticky tape.

• PRESERVED FRUITS •

There is often a need for a special pudding in an emergency, or just as a treat for a small party, and it is worth preserving a little fruit in wine or spirits for just such an occasion. A small selection of such preserves can be stored for months, and although the outlay on spirits or wine may seem high, the final cost is low if the fruit is home-grown or bought when cheap and plentiful, and each jar will provide many portions.

◆ APPLES ◆ IN WHITE WINE

8 lemons

600ml/1 pint white wine

2kg/4lb sugar

2.5kg/5lb apples

2 tbsp brandy

Peel the lemons very thinly and put the peel into a bowl. Pour on 600ml/1 pint boiling water and stir in the wine. Leave to stand for 30 minutes. Put into a pan with the juice of the lemons and the sugar. Bring slowly to the boil, stirring until the sugar has dissolved. Boil for 10 minutes and then strain. Peel and core the apples and cut them in thick slices. Put into a pan with the strained liquid and simmer until the fruit is soft but not broken. Stir in the brandy and pour the mixture into hot preserving or screw-top jars. Seal tightly.

◆ TIPSY APRICOTS ◆

250g/8oz dried apricots

600ml/1 pint boiling water

500g/1lb sugar

150ml/¼ pint cold water

200ml/6fl. oz gin or brandy

Put the apricots in a bowl with the boiling water and leave to soak overnight. Drain the fruit and chop it roughly. Put the sugar and cold water in a pan and dissolve the sugar over low heat. Add the apricots and bring to the boil. Simmer for 15 minutes. Leave for 2 hours until completely cold. Stir in the gin or brandy and store in small screw-top jars. Use as a pudding with cream, as a sauce for ices or puddings, or as a filling for pastry cases.

BRANDIED ◆ CHERRIES ◆

Morello cherries

sugar

brandy

Wash the cherries and dry them well. Trim the stalks, leaving about 12mm/½in on each fruit. Weigh the prepared fruit and allow 625g/1¼lb sugar to each 500g/1lb fruit. Mix the fruit and sugar in preserving jars and fill up with brandy. Seal tightly and leave for 3 months. Serve in small portions in wine glasses, giving each person some fruit and liquid. Eat the cherries first and then drink the resulting liqueur.

◆ FRUIT MEDLEY ◆

strawberries, raspberries, plums, peaches

brandy

This is a simple variation of Rum Raisins (p. 151) which is useful for many cooking purposes. Just use a few late strawberries or raspberries, one or two small peaches, and a handful of Victoria plums and put them in a preserving jar. Top up with brandy, seal and keep in the refrigerator. Add odd pieces of fruit when you have them. Use the fruit and liquid to add to fruit salads or to serve with ice cream.

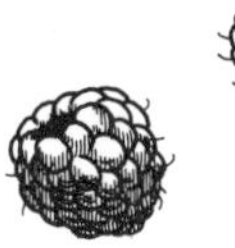

• BRANDIED PEACHES •

3kg/6lb small ripe peaches

2kg/4lb sugar

1.5 litres/2½ pints water

brandy

Dip the peaches in hot water one at a time, and rub off the 'fur' with a clean cloth. Put the sugar and water into a pan and dissolve the sugar slowly over low heat. Bring to the boil and boil for 10 minutes without stirring. Add the peaches to the syrup a few at a time and simmer for about 5 minutes until tender. Lift out the peaches with a perforated spoon and pack into hot preserving jars. Simmer the syrup until thick. Cool and measure the syrup and take an equal quantity of brandy. Bring the syrup and brandy to the boil and fill the jars. Seal tightly.

• PEARS & CHERRIES • IN WHITE WINE

2kg/4lb eating pears

500g/1lb Morello cherries

600ml/1 pint white wine

600ml/1 pint water

1kg/2lb sugar

small piece of cinnamon stick

Peel and core the pears and cut them in half. Stone the cherries. Put the wine, water, sugar, and cinnamon stick in a pan and bring slowly to the boil, stirring well to dissolve the sugar. Add the cherries and simmer until the cherries are almost tender and the syrup is thick. Add the pear halves, bring to the boil and then simmer for about 3 minutes until the pears are just tender but unbroken. Put into hot preserving jars, covering the fruit completely with syrup, and seal tightly. If preferred, white port may be used instead of white wine.

• PEARS • IN RED WINE

3kg/6lb cooking pears

500g/1lb sugar

red wine

water

Use small unripe pears, peel them and leave them whole with the stalks on. Pack into large preserving jars and divide the sugar between the jars. Half-fill the bottles with wine and top up with water. Put on the tops and stand the jars in the oven. Leave for 3 hours at 130°C/250°F/gas mark ½. Remove from oven and screw on tops tightly. Store in a cool dry place. To serve, drain the liquid from the pears and simmer the liquid until it forms a thin syrup. Pour over the pears and serve cold.

• BRANDIED PEARS •

3kg/6lb cooking pears

3 lemons

2kg/4lb sugar

8 cloves

1 cinnamon stick

6 tbsp brandy

Peel the pears, cut them in quarters and remove the cores. Grate the rind from the lemons and squeeze out the juice. Arrange layers of pears, sugar and lemon rind in a large bowl and sprinkle the lemon juice on top. Cover and leave overnight. Pur the pears into a large casserole and cover them with the juices which have formed during the night. Add cloves and cinnamon, cover and bake at 140°C/275°F/gas mark 1 for 6 hours until the pears are tender and golden. Leave in the casserole to cool and then stir in the brandy. Take out the cloves and cinnamon. Put the pears into preserving jars and seal tightly.

Pears in Port

Substitute 150ml/¼ pint port for the brandy.

• SQUIFFY PRUNES •

prunes

port or brandy

Use high quality tender sweet prunes. Pack them into small screw-top jars, such as honey jars. Fill up with port or brandy, making sure the liquid fills all the spaces in the jars. Seal tightly and keep for at least 6 weeks. The prunes become plump and make a delicious dessert served in the liquid.

• RUM RAISINS •

250g/8oz caster sugar

150ml/¼ pint water

250g/8oz seedless raisins

6 tsp rum

Put the sugar and water into a saucepan and dissolve the sugar over low heat. Bring gently to the boil and stir in the raisins. Simmer for 15 minutes. Cool for 2 hours and then stir in the rum. Store in small screw-top jars. Use as a sauce for ices or puddings.

EVERLASTING • RUMPOT •

1 bottle light or dark rum

granulated sugar

strawberries, cherries, apricots, raspberries, plums, redcurrants, peaches, grapes, melon

Use a selection of fruit, but avoid citrus fruit, apples, bananas and pears. The fruit must be sound, whole and ripe, and it is possible to use just a few choice fruits at a time as they come into season, rather than overloading the rumpot with fruit. Use a large stone crock or glass preserving jar. Wipe the fruit gently and do not peel or stone. Melon is the only fruit which should be peeled, seeded and cut into large chunks. Put the fruit into a large stone crock with its own weight of sugar and cover with the rum. Cover tightly and keep in a cool place. Start the rumpot with strawberries and cherries and add fruit as it comes into season. Keep the completed rumpot for at least 3 months before using. Use the fruit and its syrup in small quantities to eat with cream or yogurt. Some of the fruit and liquid may be added to fresh oranges, apples and nuts to make a fruit salad. The fruit can be drained and put into a pastry or sponge flan case to serve with cream. For a special pudding, take the top off a melon and scoop out the seeds. Pierce the flesh with a knitting needle and then fill the melon with fruit and syrup and serve chilled. Although the initial expenditure on rum may seem excessive, the rumpot will produce dozens of portions of delicious puddings.

DAIRY FOODS

• EGG PRESERVING •

The main function of all methods of egg preserving is to prevent the entry and growth of spoilage micro-organisms. Chickens' eggs are naturally well protected by their calcite shells and cuticle coating as they may have to keep 16 to 17 days after laying before brooding begins. Each egg is perforated with as many as 17,000 pores for exchange of gases and water by the embryo chick, but on laying the cuticle covers the majority of these. As the egg ages or is damaged these pores become exposed allowing free exchange of water and gases and penetration by micro-organisms. The action of all preservation methods for eggs in the shell is the blocking of the pores to prevent dehydration and infection.

Preservation in the shell. The eggs must be fresh and clean. New-laid eggs should stand for 18–24 hours before treating but eggs for preserving should not be more than 3 days old. Never wash eggs as this damages the cuticle – some stains can be removed with a warm cloth.

Water-glass (sodium silicate). Use 500g/1lb water-glass dissolved in 4 litres/8 pints of boiled water. Cool and pour into the preserving jar or bucket. Place the eggs pointed end downwards as this keeps the yolk suspended in the centre of the egg away from the shell. This amount of

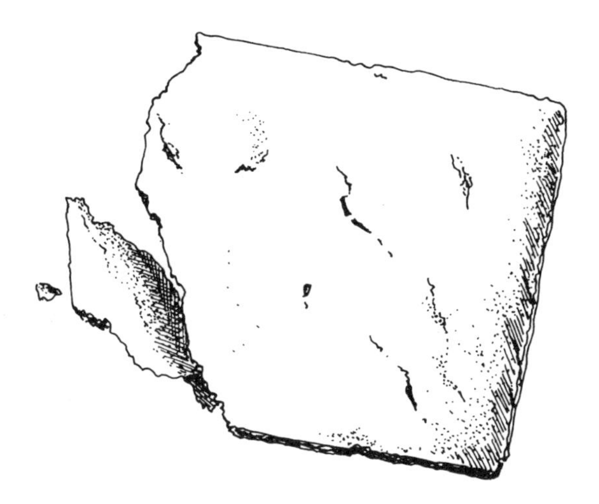

water-glass should be sufficient for about 50 eggs, but ensure that all the eggs are completely covered. Store covered in a cool (15°C/60°F) place.

Dry preserving. Coat the shells thoroughly and store pointed end downwards in a bowl or basket. For this method great care must be taken to cover the shells completely.

(i) Grease with white vaseline, lard or butter. Roll the eggs carefully in the fat in the palm of the hands. Store in bran if possible.

(ii) Cream 175g/6oz borax and 250g/8oz lard together. Coat by hand with this mixture and store in a bowl or basket.

Keeping time for preserved eggs. Keep for 4–6 months for eating, 10 months for cooking. Always prick the shells of preserved eggs before boiling or they will crack as the pores are sealed.

Freezing eggs. Eggs crack and become 'gluey' if frozen in their shells. Break them and either whisk lightly or separate and freeze yolks and whites individually. Freeze in small usable quantities. Add salt or sugar to yolks and whole egg to prevent coagulation (½ tsp salt/1 tsp caster sugar for each 2 yolks or 1 egg) and label for savoury or sweet dishes.

• BUTTER-MAKING •

It is not difficult to make butter at home, but a lot of cream is needed, so butter-making is not an economy, even if you have your own cow or goat. The cream from 10–15 litres/ 2½–3 gallons of milk is needed to make 500g/1lb butter. Butter-making complements cheese-making, as the skim milk and buttermilk left when the cream has been skimmed off can be used for cheeses. Few people will want to make large quantities of butter and will not need to separate their own cream, or buy equipment. It is however perfectly possible to make small quantities of butter for domestic use, with an electric mixer or blender.

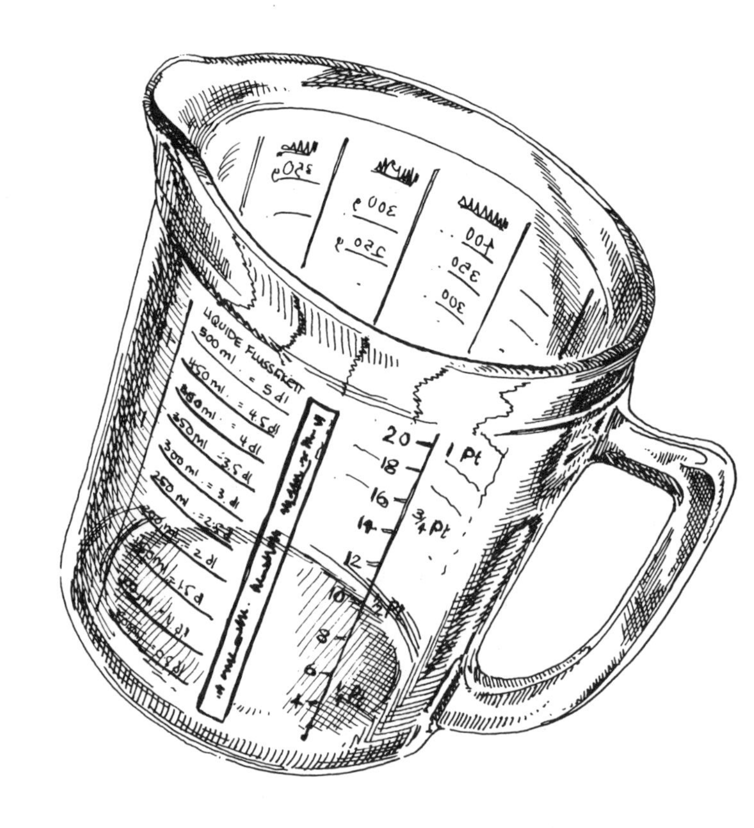

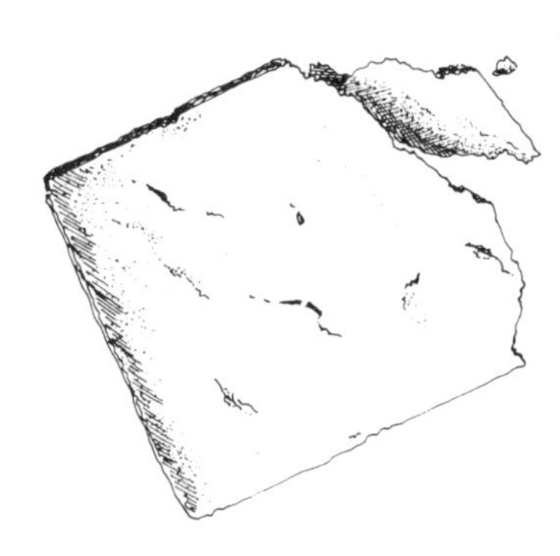

Electric mixer method. Use 4-day-old double cream and put it into the mixing bowl. Use the heavy beater on the mixer, and run it on low speed. The cream will whip and then become very thick and curdled looking. As soon as this happens, reduce the speed to the lowest setting and continue until the butter forms one lump. Drain off the buttermilk. Wash the butter in cold water until the water remains clear. Squeeze well to extract water and then add salt to taste before patting the butter into shape. 600ml/1 pint cream is a practical quantity to deal with.

Electric blender method. Blend the cream on high speed for 1 minute. Drain off the buttermilk. Put cold water into the blender goblet, and blend with the butter for a few seconds. Drain off the water and squeeze the butter to extract moisture. Salt to taste before patting into shape. For this method, 600ml/1 pint cream is also a practical quantity to use.

• YOGURT •

Yogurt may be made at home with fresh whole milk, skimmed milk, or either of these with dried milk added. When cultures are added to the milk, lactic acid is produced. A special yogurt 'starter' may be purchased but commercial natural yogurt is easy to use. This should not be pasteurised and must be fresh stock – old stock may be weak and unfit to start your new batch of yogurt. If it takes more than 6 hours to set new yogurt, it means that the 'starter' is weak and must be replaced. Your own yogurt may be used for this, or some more commercial yogurt.

Various types of milk give a different yogurt, so it is possible to produce rich yogurt, or thick or thin varieties. Raw fresh milk, pasteurised, or homogenised milk (or a mixture of all three) boiled for 5 minutes and cooled to 46°C/115°F will produce average yogurt similar to the commercial variety widely available. Sterilised milk will produce a thin yogurt and so will UHT (long life) milk, unless boiled and cooled as above. If any of these milks are partly or wholly skimmed, they will produce a thicker yogurt. The addition of 125g/4oz dried skimmed milk powder to 600ml/1 pint boiled or sterilised milk will give thick yogurt. Dried skimmed milk may be used with boiled and cooled water (75g/3oz powder to 600ml/1 pint water) or it may be used on its own. It is obviously very convenient to be able to use UHT and powdered milk in this way as it ensures a regular supply of yogurt even if fresh supplies fail in bad weather. All yogurt is made with a natural flavour which may be flavoured with fresh or thawed frozen fruit.

Yogurt-making is simple. The milk and starter may be put into a container in a warmed blanket and left to rest in an airing cupboard or on a heater which gives an even temperature. An electric yogurt-making machine may be used, or a wide-necked Thermos flask. The important factor is that the surrounding temperature should be slightly warmer than the yogurt itself. All equipment must of course be clean and sterile.

Use yogurt as a simple dessert or breakfast snack, adding honey, sugar or fruit. Alternatively, use instead of cream in savoury goulashes and sweet mousses and cheesecakes.

• SIMPLE YOGURT •

600ml/1 pint milk

2 tbsp natural yogurt or starter

Boil the milk for 5 minutes and cool to 46°C/115°F. Put a little into a sterile bowl and whisk in the yogurt or starter. Whisk in the remaining milk to blend it well. Put into a sterile container with a lid, wrap in a warmed blanket and keep at a temperature of 43°C/110°F for 3–4 hours. When a curd has formed, cool the container at room temperature and then refrigerate for 10–12 hours. The milk mixture may be left up to 6 hours to incubate, but should not be over-incubated. Use any of the suggested milks or a mixture of them. Add flavouring to taste, or, if preferred, use unsweetened.

• CHEESE-MAKING •

Hard cheeses are not easy to make at home, as methods can be complicated and the finished cheeses are large and heavy to handle. A 4kg/8lb cheese for instance, may mean dealing with 28 litres/8 gallons milk which means large equipment and heavy weights to lift. It is better to make one or two soft cheeses, starting with the traditional cheese made from naturally soured milk. It is then possible to make a few simple cheeses in which the milk is activated by a cheese rennet.

• CREAM CHEESE •

1 litre/2 pints double cream

salt

Put a wet double layer of close-textured cloth or muslin into a soup plate. Put in the cream and leave it to stand for 24 hours until thick. Turn the cheese and add a pinch of salt. Leave for 12 hours. Put into a dry cloth, salt the other side and leave for another day. This cheese used to be chilled for service, and was often covered with nut leaves or nettle leaves.

• DESSERT CHEESE •

1 litre/2 pints milk

600ml/1 pint double cream

2 tbsp junket rennet

1 egg yolk

sugar

Warm the milk and cream gently to 80°C/175°F and cool to 27°C/80°F. Stir in the rennet. Cover and leave to stand until it forms a firm jelly which may take 8–12 hours. Break up the jelly and put into a sieve to let the whey drain off. Put the curd into a dish with the beaten egg yolk and sugar to taste, and mix well. Serve with fresh cream and more sugar.

• POT CHEESE •

sour milk

cream or butter (optional)

herbs (optional)

salt

Leave the milk in a clean, cool place until it sours naturally and forms a curd. Tip it carefully into a piece of muslin or close-textured cloth (a piece of worn sheeting is good for this purpose). Tie the ends of the cloth together to form a bag and hang it over a sink, or put it in a sieve over a basin, so that the curd can drain away. From time to time, move the thick curd with a fork to release the whey. When the curd is firm, scrape it out of the cloth. Add a little cream or butter and some chopped herbs if liked, and salt to taste. Serve this cheese while it is fresh, as it can quickly become very bitter.

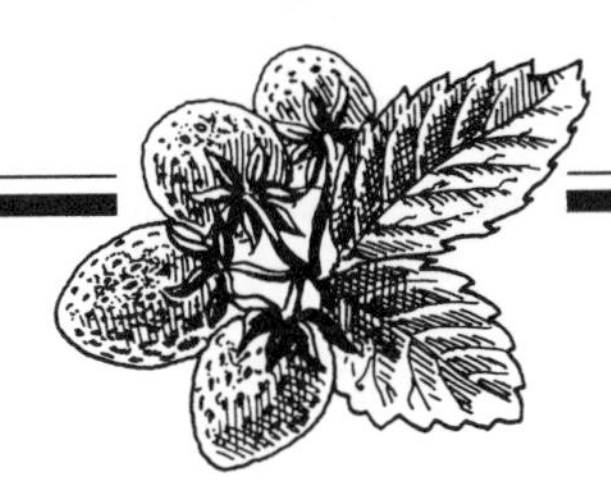

INDEX